THE COMMONWEALTH AND INTERNATIONAL LIBRARY

HIGHER MATHEMATICS FOR SCIENTISTS AND ENGINEERS

General Editor   F. M. ARSCOTT

NUMERICAL ANALYSIS

NUMERICAL ANALYSIS

I. M. KHABAZA

Reader in Computing Science,
Queen Mary College, University of London

PERGAMON PRESS

OXFORD · LONDON · EDINBURGH · NEW YORK
PARIS · FRANKFURT

Pergamon Press Ltd., Headington Hill Hall, Oxford
4 & 5 Fitzroy Square, London W.1

Pergamon Press (Scotland) Ltd., 2 & 3 Teviot Place, Edinburgh 1

Pergamon Press Inc., 122 East 55th Street, New York 22, N.Y.

Pergamon Press GmbH, Kaiserstrasse 75, Frankfurt-am-Main

Federal Publications Ltd., Times House, River Valley Rd., Singapore

Samcax Book Services Ltd., Queensway, P.O. Box 2720, Nairobi, Kenya

First edition 1965

Library of Congress Catalog Card No. 64–66143

Set in 10 on 12 pt Times
and Printed by Adlard & Son Ltd. Dorking

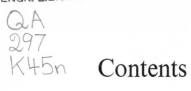

Contents

CONTENTS

Foreword

IN RECENT years, a phenomenal increase has taken place in the amount of mathematical knowledge required by practising engineers and scientists, which in turn demands an unprecedented widening and deepening of their mathematical education. Topics which, only a decade ago, were considered too advanced and abstruse for any but mathematical specialists now form optional parts, at least, of the undergraduate curriculum at many Universities. There is every reason to think that this trend will continue, so that before long all graduate scientists and engineers will be expected to have, not as a luxury but an essential part of their mathematical equipment, some knowledge of the material which we have sought to cover in the five volumes on "Higher Mathematics for Engineers and Scientists" in the Commonwealth Library series.

In broad terms, these are written for the student who has already completed a leisurely two-year course, or a concentrated one-year course, of non-specialist mathematics at University level, reaching a standard which in terms of examinations is approximately that of the Ancillary Mathematics for Students of Physics and Chemistry, or the B.Sc. (Engineering) Part II, of London University. The two volumes *Elementary Mathematics for Scientists and Engineers* by M. D. Hatton, and *Further Mathematics for Scientists and Engineers* by R. Tearse in the Commonwealth Library series, cover this preliminary material completely, and the five volumes in the Higher Mathematics set rest upon them as their foundation.

To determine the scope of these volumes, and the distribution of material between them, my colleagues and I have taken as our guide the (optional) mathematics section in the syllabus for the

B.Sc. (Engineering) Part III degree of London University. The five books in this set cover, respectively, all the material in sections I–V of that examination. We have not, however, hesitated to go some distance beyond these limits from time to time in order to include other topics of special importance, and hope that as a result the books will meet the needs of most non-mathematical students the world over. This has been particularly the case in the volume on Numerical Analysis, where in anticipation of future needs much greater stress has been laid on the use of digital computers than the examination requirements alone would justify. At the same time, we have excluded (sometimes very reluctantly) any major matter which seems unlikely to form, in the foreseeable future, part of an undergraduate course for non-mathematicians.

Although the five volumes may be used independently, they have been written in collaboration and are intended to be complementary. Most of the material on differential equations, for instance, is to be found in *Differential Equations and Special Functions*, but the techniques of numerical solution and the integral-transform method are dealt with respectively in *Numerical Analysis* and *Complex Variable and Integral Transforms*, while some of the underlying calculus is covered in *Advanced Calculus and Vector Field Theory*.

Intended as they are for scientists and engineers, the mathematical presentation of material in these volumes has been deliberately kept at a somewhat unsophisticated level. Nevertheless, we have sought to maintain a reasonable standard of mathematical rigour, and hope that even the specialist mathematician may find the books useful as introductory material before passing on to stronger meat.

The thanks of the authors are due to the members of the Board of Editors of the Engineering division of the Commonwealth Library for many helpful comments and criticisms, both in the planning stages and subsequently; other acknowledgements are in the individual volumes. Collectively, we hope that this project

will do something to fill the gap between the resources of mathematics and the needs of natural science and engineering, which is
already large and appears to be growing with alarming speed.

Department of Mathematics,   F. M. ARSCOTT
Battersea College of Technology,
London S.W.11

*Commonwealth Library "Higher Mathematics for Engineers
and Scientists"*

Advanced Calculus and Vector Field Theory, *by* KATHLEEN M. URWIN
Differential Equations and Special Functions, *by* F. M. ARSCOTT
Complex Variable and Integral Transforms, *by* H. V. MILLER
Advanced Dynamics, *by* W. DAVIDSON
Numerical Analysis, *by* I. M. KHABAZA

Preface

THIS book has developed from a course given to Engineering students of Battersea College of Technology. Although the limitations and pitfalls of the various methods are indicated, the emphasis at this elementary level is on the development of a feeling for the subject rather than on rigour. Methods suitable for digital computers are emphasized, but some desk computations are also described. Although desk machines are hardly used today in practical problems, they are discussed because they may be useful in a course for training beginners in the computing art; they are convenient for acquiring at first hand a feeling for numerical techniques and an appreciation of the reason why some methods are more sound or more stable than others.

No references are mentioned in this elementary introduction to the subject. There is instead a short list of books suggested for further reading. The last two items in the list contain detailed references and bibliographies. I would like to take this opportunity to acknowledge my debt to all the books mentioned in this list.

I owe a great debt to Dr. J. C. P. Miller of the Cambridge Mathematical Laboratory, who made extensive suggestions for improving the first draft of this book; but I remain responsible for any defects.

My thanks are due to the University of London Computer Unit, whose Mercury Computer I used for working out many of the tables and examples in this book; to Prof. F. M. Arscott, head of the Department of Mathematics at Battersea College, who suggested this book; to the Senate of the University of London and the Governing body of Battersea College of Technology for permission to use examination questions (marked U.L. and B.C.

respectively); to the staff of the Pergamon Press for their patience and courteous help.

I. M. KHABAZA

SUGGESTIONS FOR FURTHER READING

BUCKINGHAM, R. A. (1962) *Numerical Methods*, Pitman, London.
HILDEBRAND, F. B. (1956) *Introduction to Numerical Analysis*, McGraw Hill, New York.
National Physical Laboratory (1961) *Modern Computing Methods*, H.M.S.O.
TODD, J. (1962) *Survey of Numerical Analysis*, McGraw Hill, New York.

The last two books contain detailed references and bibliographies.

NOTE: The symbol $\sim$ is used throughout this book to mean "is approximately equal to".

CHAPTER 1

Digital Computers

THE DESIGN and development of numerical methods often depend on the computational aids available. Analogue computers and similar devices are an important aid, but they are not considered in this book. The use of desk machines is discussed in the next chapter. In this chapter we consider digital computers; as they are being used increasingly, an understanding of the general way in which they work is necessary in order to understand the motivation behind many numerical techniques. The field of digital computers is expanding rapidly, so that even the brief description given in this chapter may need modification in a few years' time.

1.1 General Description

The importance of digital computers in numerical work is due to their speed, which is approaching a million arithmetical operations a second, and to their ability to carry out a complicated set of instructions including tests for alternative courses of action.

Most numerical work consists of some data X which determines a solution Y of the problem. For example, the data X may be the matrix of the coefficients and constants of a set of linear equations, the solution Y being a set of the values of the unknowns which satisfies the equations; or the data X may be the coefficients of a polynomial, the solution Y being the set of zeros of the polynomial. A program is a method of solution; it may be thought of as a function P so that $Y = P(X)$. In this relation X is the input, Y the output, and P the program.

A digital computer will then have an input device to read the program P and the data X, and an output device to punch or print out the results Y. It must also have a control unit which carries out the instructions of the program, an arithmetic unit to carry out the arithmetic operations, and a store for the data, the program and intermediate results.

1.2 Input and Output Systems

Programs and numerical data are punched on paper tape or cards for input to the computer. The output from the computer can also be on paper tape or cards. Paper tapes are $\frac{1}{2}$ inch to 1 inch wide; a character occupies 5 to 8 hole positions or channels across the tape, packed at about 10 characters per inch. A card is about 3 inches by 8 inches, usually divided into 80 columns and each column has 12 hole positions; a character usually occupies a column, but there are various other ways of packing information on a card. Cards and card equipment are usually more bulky than paper tape and tape equipment, but cards are easier to prepare and edit and more suitable for large amounts of data.

Input devices are often faster than output because input usually involves photo-electric sensing and output involves mechanical punching. Output devices can slow down computers considerably; this defect can be overcome by the use of line printers which can print, rather than punch, about 1000 characters per second when used efficiently. Another way of overcoming this difficulty is to record the output on a magnetic tape which is then fed into off-line equipments which do not hold up the central computer. Magnetic tapes can be used similarly to speed up input.

Numbers are usually stored inside the machine in binary form, but it is not practical to prepare the data in this form. Usually a program is used which will convert data punched in decimal form to the binary or appropriate machine form. The reverse procedure is followed for output: a program converts the numerical results from the binary machine form to decimal form.

A similar principle is used for program input. The machine form of instructions is often in binary code, but it takes some time to be able to read and write machine code with ease and it is easy to make mistakes and difficult to spot them. A master program, the input routine, is used instead. This program enables the programmer to write instructions in a more intelligible form. Input routines can be more or less elaborate. The simplest form will do no more than convert instructions in convenient decimal code to the binary machine code. Next, an input routine may enable the programmer to refer to numbers and instructions by convenient labels attached to them; otherwise the programmer would have to keep a record of the position of every instruction or number. More elaborate forms of input routines, called automatic languages, enable the programmer to write instructions in the form of algebraic formulae and the input routine will compile in machine form the instructions necessary to evaluate these formulae. Elaborate input routines can take up a lot of space and time in compiling a program, and the program compiled by input routines is not always as efficient as it could be if it were written directly in machine code by an experienced programmer; none the less automatic languages justify themselves by making programming easy and the resources of computers accessible to a wide class of users who do not need to be expert programmers; the disadvantages of automatic languages decrease as computers become larger and faster.

1.3 Storage Systems

Information such as numerical data or programs can be stored in a computer in the form of acoustic pulses travelling at the speed of sound in delay lines. It can also be stored in the form of a train of electric pulses travelling at the speed of light. In a stationary form information can be stored in a series of flip-flops with high and low potentials indicating their state. In a magnetic form it can be stored in cores made of ferrite materials, the direction of magnetization indicating the state of the core; this is an efficient and, at the moment, the most common form of storage.

Information can be stored on fast rotating drums, the surface of the drum being coated with iron oxide; the magnetic state of small areas on the surface of the drum indicates the binary information stored. Magnetic tapes may also be used for storage; they are usually slower than magnetic drums, but suitable for storing large amounts of data such as occur in commercial and industrial work; magnetic tapes are also used for input and output systems.

According to the time of access, storage systems are classified into slow and fast stores. The access time to slow devices, such as drums, is measured in milliseconds; whereas fast storage systems, such as core stores, have an access time of the order of a microsecond. Large slow stores can be combined effectively with small fast stores; data or programs are divided into blocks and called from the slow backing stores to the fast immediate access store whenever needed; the computation is speeded up considerably by dividing the program into blocks of suitable size to fit in the fast stores so that blocks are changed as little as possible; but the need for this is getting less frequent as fast stores are becoming larger. In this context fast stores are sometimes called immediate access stores; as instructions and data in the backing stores cannot be processed, they have first to be transferred to the fast store. The unit of storage is called a *word*; a word has about 32 to 48 binary digits or bits and it is usually of a fixed length. A backing store may have room for 100,000 words while the fast store may have room for 16,000 words.

1.4 Arithmetic Unit

Numbers are usually stored in binary form. High potential might indicate 1 and low potential 0. The two magnetic states are suitable to represent 0 and 1. A pulse could represent 1 while an absence of pulse would indicate 0. The number of binary digits in a word is called *word length*. Numbers are usually recorded to the full length of the word. Thus, for example, if the word length is 5, the number 3 is represented by 00011. Counting from right to left, the first digit represents $2^0 = 1$, the second digit represents

$2^1 = 2$, the third digit represents $2^2 = 4$, and so on. Thus 10111 represents:

$$1 \times 2^4 + 0 \times 2^3 + 1 \times 2^2 + 1 \times 2^1 + 1 \times 2^0$$
$$= 16 + 4 + 2 + 1 = 23.$$

For convenience we consider a word length of 5 bits; the same principles apply to a word length of 32 or 40 or 48 bits. A 5-bit word can represent numbers between 0 and 31. Thus the binary number 11111 represents 31. A simple convention can be used to represent negative numbers. We observe that

$$11111 + 00001 = 00000$$

because there is no room in a 5-bit word for the carry from the 5th bit. It appears that in 5-bit binary arithmetic 31 is equivalent to -1. This is equivalent to assuming that the 5th bit represents -2^4 instead of $+2^4$ while the remaining bits represent the same positive numbers as before. This assumption is called the sign convention. In the sign convention a 5-bit word can represent numbers in the range -16 (which is 10000 in binary) to $+15$ (which is 01111 in binary). Thus the addition

$$\begin{array}{r} 11101 \\ +00101 \\ \hline 00010 \end{array}$$

means, in the sign convention: $-3 + 5 = 2$, whereas in the unsigned convention it means: $29 + 5 = 2$ (modulo 32), i.e. the digit corresponding to 32 is lost because it is in the 6th place and the word has room for 5 places only. When a digit is thus lost the term *overflow* is used. In the sign convention a number is positive if the most significant digit is 0 and negative if the most significant digit is 1. The sign convention makes it possible to handle negative numbers on a machine which cannot distinguish between positive and negative numbers in arithmetic operations.

This form of arithmetic is called *fixed point arithmetic*. Numbers are treated as integers, and fractions are dealt with by scaling factors. For example, in a 32-bit word one may have a scaling factor of 2^{-20}. This means that a "binary point" is imagined between the 20th and 21st bit. In decimal arithmetic this is equivalent to about 6 decimal places after the decimal point and 3 before it. This form of arithmetic has the disadvantage that scaling factors have to be kept in mind constantly and a lot of programming is necessary when numbers which vary considerably in magnitude occur in the same calculation.

An alternative form of arithmetic on digital computers, which avoids this difficulty, is called *floating-point arithmetic*. A number x is represented in the form of a pair of numbers (a, b) where b is a fraction in the range:

$$\tfrac{1}{2} < b < 1 \qquad \text{or} \qquad -1 < b < -\tfrac{1}{2}$$

and a is an integer such that $x = 2^a \times b$. We say that a is the exponent and b the argument. Thus in a 32-bit word 8 bits may be reserved for the exponent and 24 bits for the argument. The arithmetic on the 8-bit exponent would be fixed point in the sign convention allowing for numbers as large as $2^{2^7-1} = 2^{127}$ and as small as $2^{-2^7} = 2^{-128}$. The arithmetic on the argument part of the number is also fixed point in the sign convention, but it has a constant scaling factor of 2^{-23}. This is equivalent to saying that the most significant bit represents $-2^0 = -1$, the next significant bit represents 2^{-1}, the next 2^{-2}, and so on.

The accumulator is an important part of the arithmetic unit. Most arithmetic operations boil down to adding into the accumulator or subtracting from it. Under the sign convention it is possible to get a negative number as a result of adding two large positive numbers. This can be avoided by scaling down before adding. But usually the accumulator has an extra guard digit which indicates overflow when this happens. In floating-point arithmetic the result is scaled down if there is an overflow in the addition of the argument, and the exponent is correspondingly increased. If there is an overflow in the exponent, which happens only when the content of the accumulator is extremely large, then the machine is made to stop or to indicate overflow of the exponent.

On some machines the basic form of arithmetic is fixed point arithmetic. Floating-point arithmetic can still be carried out by programming, but then it is much slower than the basic fixed point arithmetic. On other machines the basic form of arithmetic is floating-point; other machines and possibly most future ones have both.

It often happens that the exact result of an arithmetic operation requires more digits than there are in a single word. Thus the exact product of two numbers of n digits each contains $2n$ digits. The direct formation of this exact product is usually possible only if the accumulator is double length. In this case the product is rounded off to a single length number by adding 1 to the most significant digit of the least significant half and dropping the resulting least significant half. In machines where the accumulator is single length, round off is not so easy or accurate; it is still possible to carry out double length arithmetic by programming but this is much slower. (See § 2.4.)

1.5 The Control Unit

In a digital computer an instruction is stored in a form which is physically the same as a number. It consists of a word of, say, 32 digits; but the 32 digits are divided into groups of smaller numbers which specify the instruction. The most general instruction has the form (R, a, b, c, d) where R is a number which specifies the function; this could be addition, multiplication, ... , according to the value of R. The numbers a, b specify the addresses of the operands, i.e. the numbers to be added or multiplied. The number c is the address where the result of the operation is to be stored. The number d is the address of the next instruction to be obeyed. Thus in a 32-bit word, 4 bits may be reserved for the function R allowing $2^4 = 16$ different operations, and 7 bits for each of the addresses a, b, c, d allowing $2^7 = 128$ addressable locations in the fast store. This is a small fast store, and one or more of the numbers b, c, d is dispensed with allowing a larger addressable fast store. Thus if it is assumed that instructions are usually obeyed in the order in which they are stored, d may be dropped and the machine is said to be sequential. If it is assumed that one of the two numbers to be operated on is already in the accumulator, then b may be dropped. Finally, if the result of the operation is left in the accumulator, then c may be dropped. An instruction of the form (R, a, b, c, d) is said to be of the $3 + 1$ address type, the form (R, a, b, c) is a 3-address type, . . . , the form (R, a) is a one-address type. Most modern machines have one-address instructions, these instructions have more digits available for the address a and the function R, allowing a larger fast store and a larger variety of functions; one-address instructions make it possible to have shorter words for instructions and are faster to obey than the multi-address instructions. More recent machines have some instructions of the zero-address type, only R is specified; the operands are taken from successive positions of a nested pop-up store.

An instruction about to be obeyed is usually transferred from the fast store to the control unit where it is decoded. The operation is selected according to the number R and the operands are fetched from the fast store to the arithmetic unit according to the addresses a, b, c; when the operation is completed the next instruction is transferred to the control unit according to the address d or, in a sequential machine, from the address after that of the instruction just terminated. Thus the fast store contains two types of words: instructions and numbers. Often these two types are physically similar, but for programming convenience they may be separated; thus in a fast store of 1000 words the first 500 words may be reserved for instructions and the remaining 500 words for numbers.

Beside arithmetic instructions there are jump instructions. In an instruction of the form (R, a) the function specified by R may cause the control to be transferred to the instruction in location a instead of proceeding to the next location. Some jump instructions are conditional, the control is transferred to location a only if, say, the accumulator is positive. The conditional jump instructions make it possible for a group of instructions, a *loop*, to be obeyed several times. A *cycle* is a loop which is obeyed a specified number of times. Thus, to evaluate a polynomial of order n, a group of instructions may be obeyed exactly n times. A second type of loop is iterative; the loop is repeated as many times as is necessary to achieve a certain accuracy or a certain aim. Thus in Newton's formula

$$x_{n+1} = x_n - \frac{f(x_n)}{f'(x_n)}$$

for evaluating a zero of the function $f(x)$, we test if $f(x_n)$ is numerically less than a certain specified accuracy, E, e.g. $E = 0 \cdot 000,001$. If $\mid f(x_n) \mid - E$ is positive the loop for evaluating the next approximation x_{n+1} is repeated, otherwise the control is transferred to the next stage.

If we wish to add, for example, 10 numbers $x_1, x_2, \ldots, x_{10}$, the numbers would normally be stored in 10 consecutive locations, and the addition would need 10 instructions: take x_1 to the accumulator, add x_2, add x_3, ... , add x_{10}. The sum is formed in the accumulator. This obviously wastes valuable program space, and would not work for adding a different set of numbers of different length. The 9 add instructions differ only in the address of the operands, and these are 9 consecutive numbers. One way of saving space is to obey the same add instruction 9 times and alter the address part of the instruction by adding 1 to it each time. In this context we see the advantage of the fact that instructions have the same form as numbers and can be treated as such; by transferring the instruction word to the arithmetic unit and adding a suitable number to it we can alter the address part of the instruction. But this is a time-consuming form of address modification; in modern machines the address part of the instruction contains only a presumptive address. The actual address is the presumptive address added to the content of a modifying register specified in the instruction. Modifying registers are called B-lines in some machines and index registers in others, there may be 8 or 16 or more of them. An effective use of these registers is made when they are used at the same time to modify an address and to count the number of times a cycle loop is repeated.

Besides arithmetic and jump instructions a computer must have a set of instructions for reading or input, for punching or output, for transferring blocks of data or programs from the backing store to the fast store and vice versa. There are also instructions which perform various logical and shift operations on the bits of a word.

1.6 Some General Programming Concepts

We must first explain certain simplified notations used in this section. The statement or command $y = ax + b$ in a program has the following meaning: there are locations in the machine denoted by y, a, x, b, ... The statement stands for the group of machine instructions which will replace the content of location y by the product of the contents of locations a and x added to the content of location b. We observe that this notation is ambiguous in two respects. First it does not distinguish between a location and its content. Secondly the equal sign $=$ is used in a dynamic sense and not in its usual mathematical sense; in this notation $y = y + 1$ makes sense; it means: increase the content of y by 1, whereas the mathematical equation $y = y + 1$ does not make sense. In books on programming these ambiguities are removed by introducing new notations. A term such as a_r means that the number a_r is taken from consecutive locations a_0, a_1, a_2, ..., and r may be regarded as the content of a modifying register.

Consider the problem of evaluating the polynomial

$$a_0 x^n + a_1 x^{n-1} + \ldots + a_n$$

and putting this value in location y (which can be taken as the accumulator). This could be accomplished by the recurrence relation

$$y = xy + a_r$$

for $r = 1, 2, \ldots, n$, with $y = a_0$ initially. A *flow diagram* of this procedure is shown in Fig. 1. Notice how r is used as a modifier as well as a counter. The diamond-shaped box is normally used for a conditional jump.

Most programs are divided into parts called *routines*. There would be a master flow diagram showing briefly the function of each routine and the flow of the calculation into the various routines, then each routine would have a detailed flow diagram

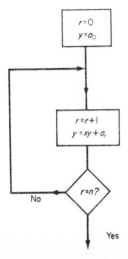

Fig. 1. Flow diagram for evaluating a polynomial.

of the type shown in the figure. Routines of this type are said to be *open routines* where the exit address is the same every time the routine is obeyed. A *closed routine* or *subroutine* is often a basic routine for evaluating an elementary function such as sine or cosine or for integration or some basic procedure which is used several times and at different places in the program. It is terminated by a jump instruction with an unspecified address; before entering a subroutine the return address or link is planted in the address part of this jump instruction. A simpler method is to end the subroutine by a modified jump instruction with 0 in the address part, and plant the return address in a modifier before entering the subroutine. Other techniques are used for storing and picking up return addresses in sub-subroutines and subroutines of higher order.

Although we are only concerned with their numerical applications, we conclude this brief description of digital computers by mentioning some of their non-numerical applications. These include: data processing, clerical work of more or less routine nature, and many aspects of automation; decision problems in the field of industrial management, military logic and mathematical logic; language translation and various problems of linguistic analysis, including the analysis of automatic languages for use on computers.

CHAPTER 2

Desk Machines
Errors in Computations

2.1 Basic Operations on Desk Machines

Familiarity with desk machines is very desirable if the reader is to grasp many of the numerical processes explained in this book. After one or two hours practice it is possible to carry out arithmetical operations on such machines faster and more accurately than with tables or a slide rule. This section deals mainly with hand machines, but some of it applies to electric desk machines also.

A desk machine has three registers: the accumulator A, the setting register S and the counter C.

Numbers can be set on S by moving levers or pressing buttons; numbers set on S can be added to the content of A by rotating an operating handle forward, and subtracted by rotating the operating handle backward. The counter C counts the number of times the operating handle has been rotated. The registers S and C may have about 6 to 10 places and A about 12 to 20 places.

A negative number on A is represented by its zero-complement. Suppose, for example, that the capacity of A is 4 places only, then -3 is represented by 9997 because if we add 3 to 9997 the result is 0000. On the same A, -30 is represented by 9970 and -308 by 9692. Thus the rule is to take the 9-complement of all but the last significant figure, for which we take the 10-complement, and to leave any zeros to the right of it unaltered.

The registers S and A can be shifted with respect to each other by moving a carriage. There is an arrow or a dot on C which we call the C-marker and which indicates the position of the carriage.

Thus if S is above the units position of A the C-marker points to 1, if S is above the tens position of A the C-marker points to 2, and so on.

To find, say, 54×4256 we first clear the registers A, S, C, and set 4256 on S. With the carriage in position 2 add S 5 times to A, then with the carriage shifted to position 1 add S 4 times to A. Now C will contain the multiplier 54, the register S will contain the multiplicand 4256, and the register A will contain the product 229824. Note that on desk machines it is better to get into the habit of multiplying from left to right, because of the possibility of short-cuts described below.

To deal with decimal fractions there are markers which can slide along the registers to mark the position of the decimal point. Thus to multiply $5 \cdot 4$ by $4 \cdot 256$ we put the decimal marker in the first position on register C, in the third position on register S, and in the 4th position on register A. Otherwise the multiplication is carried out exactly as before.

If, for example, it is required to find 829×4256, this could be done as before. Alternatively we could carry out the multiplication as follows: clear the registers A, S, and C; set the multiplicand 4256 on S; add once, subtract twice, add 3 times, and subtract once with the carriage in positions 4, 3, 2, 1 respectively; now the register C will contain the multiplier 829 and the register A will contain the required product. This is an important short-cut which saves a lot of time; it replaces the digit 8 in the multiplier by $10 - 2$ and replaces 29 by $30 - 1$.

As stated above, the register C counts the number of turns of the operating handle. In most desk machines the register C can count negative terms positively. Thus if C is cleared and the operating handle is given two negative turns the number 2 appears on C, usually with some coloured mark to warn that the number on C is -2 and not $+2$. If, however, C is not cleared first the number $...99998$ may appear on C although the original content of C was 0. Thus if a number, say 7, is on C, this could be either $+7$ or -7. There is usually some coloured indicator to show which. If the operating handle is now given $+2$ turns then the result on C will be 9 if the original number on C was +7, and 5 if the original number on C was −7.

To divide a by b, first set a on the register A as far to the left as possible. On some machines this can be done directly, on others one has to clear A first, then set a on S and add it to A. Then set b on S so that the most significant figure of b is in line with the most significant figure of a and the C marker is as far to the left as possible. Clear C, then subtract b as many times as possible without A becoming negative; a bell usually rings when b is subtracted once too often and A changes from positive to negative. Then the carriage is shifted to the next position on the right and the procedure is repeated as many times as the number of significant figures required in the quotient. The quotient will appear on C and the remainder on A. The position of the decimal point on C is the position of the C marker when the decimal points on S and A are placed in line. This process is known as the "teardown" method; it can be speeded up by subtracting once too often when the remainder is more than half the divisor and then adding instead of subtracting in the next position until A becomes positive again. This method of speeding up division is similar to that in multiplication.

Another method of evaluating a/b is to set b on S with A and C clear, and the C marker as far to the left as possible. Then add S to A as many times as possible without A exceeding a, then shift the carriage to the next less significant position and repeat. The position of the decimal point is determined as above. This method amounts to building up a in A by multiples of b in the various positions of the carriage. It is known as the "build-up" method. It is more liable to error than the tear-down method, unless a is simple (for example, $a = 1$ when finding the reciprocal of b, or a is some simple integer); otherwise we have the difficulty of comparing numbers that are not exactly alike.

A third method is to compute $1/b$ by the build-up method and multiply this reciprocal by a. This method is suitable especially when we have to evaluate

$$a_1/b, \ a_2/b, \ a_3/b, \ \ldots \ ;$$

we evaluate $1/b$ once and for all then

$$a_r/b = a_r \times (1/b).$$

Methods such as this for saving divisions are useful on desk machines and even on some digital computers where division is much longer than multiplication.

2.2 Further Techniques on Desk Machines

Suppose a, b, c are all given to 4 decimal places and it is required to find the product abc also correct to 4 decimal places. We set the decimal point in position 4 on C and S and in position 8 on A, then form $a \times b$ in A. On many machines there is a way of "back-transferring" this product on A to S; this is done with the C marker in position 5 so that only the most significant 4 decimal places are transferred to S. The product ab should be rounded off to 4 decimal places preferably before the transfer. If there is no such automatic facility for back-transfer, S has to be set by hand. Now ab is on S and this is multiplied by c in the usual manner.

To evaluate a polynomial, e.g.

$$ax^3 + bx^2 + cx + d$$

the "box" or "nest" method of §1.6 is used. Set a on S and multiply by x, then add b to A so that now A contains $ax + b$; back-transfer to S, multiply by x and add c so that now A contains $ax^2 + bx + c$; back-transfer to S, multiply by x and add d so that now A contains the required value of the polynomial. An important aspect of this procedure is its repetitive nature, it is a loop. The arrangement of the work, or strategy, is a program; one has to program a computation on desk machines just as much as on digital computers. This procedure for evaluating polynomials on desk machines is often complicated by the presence of negative numbers, but it will still work as long as the content of A is positive at each stage; in theory it should work even if A becomes negative when a string of nines on the left indicate that the number is negative, but this is troublesome in practice. Often a slight rearrangement of the arithmetic avoids this trouble.

To find the square root of a positive number a on a desk machine, we use the iterative formula

$$x_{n+1} = \frac{1}{2}\left(x_n + \frac{a}{x_n}\right)$$

We start with a reasonable guess x_1 and set it on S, multiply it by a suitable number so that the product on A is as near to Aa as possible (division by the build-up method). Then C contains a/x_1; this number will be nearly the same as x_1 so that one can calculate mentally the mean $\frac{1}{2}(x_1 + a/x_1)$; this mean is taken as x_2 and set on S, as S already contains x_1 this requires only few alterations. Now we clear A and C, and repeat the process to find x_3 from x_2, and so on. This is a rapidly convergent procedure; x_3 will be accurate to about 6 significant figures; the order of convergence of such procedures is discussed in §4.4.

We explain a second method of finding square roots which is much longer than the first one though it can be speeded up considerably with practice and short-cuts; the method could form the basis for the design of an automatic square root on a digital computer. To find the square root of a number (e.g. 365·3) to m significant figures (e.g. to 6 significant figures, which in this case means to 4 decimal places), set the number on A so that the most significant pair of digits are in position $2m$ and $2m - 1$. (The digits are paired from the decimal point thus: 03 65· 30. In this case 03 will be in position 12 and 11.) If in this position the decimal point of the number on A is in position $2p$, put the decimal point on S and C in position p. (In the example considered, the decimal point will be in position 8 on A and 4 on S and C.) S and C are cleared.

(1) Move the carriage until the counter marker is in position m.
(2) Set up 1 in the mth position of S and subtract once.
(3) Increase the number in S by 2 units in the mth position by advancing the setting lever or button by 2 units, and subtract again.

(4) Repeat this last step, i.e. increase by 2 and subtract again, until A becomes negative when usually a bell rings. Then cancel the last subtraction by giving the operating handle a forward turn. At this point check that the number in S is odd. If the number on S is 9 or 49 then to advance by 2 we change the number on S to 11 or 51 so that in such cases two levers have to be altered.

(5) Decrease the number in S by 1. Check that at this stage S is even and the number on S is twice the number on C.

Now steps 1 to 5 are repeated with m replaced by $m - 1$, then by $m - 2$, down to 1. The square root appears on C. If the remainder in A is more than half the number on S, i.e. more than the number on C, then another unit is added to the least significant digit of C to round off. The justification of this procedure is as follows: we are subtracting 1, 3, 5, ..., $2n - 1$ where n is an integer between 0 and 9. The total amount subtracted is

$$1 + 3 + 5 + \ldots + 2n - 1 = n^2.$$

Also, in step 5, we subtract 1 from $2n + 1$ to obtain $2n$. Suppose that we wish to find the square root of

$$(10^{m-1} n_m + \ldots + 100n_3 + 10n_2 + n_1)^2$$

where $n_1, n_2, \ldots$, are integers between 0 and 9. Then the effect of the general steps 1 to 5 is to subtract

$$2(10^{m-1}n_m + \ldots + 10^r n_{r+1}) n_r + 10^{r-1}n_r^2$$

This procedure can be speeded up by switching to ordinary division when half the required number of significant figures has been obtained. Another way of speeding up this procedure is to replace the general steps 2 to 5 as follows: estimate n_r which is approximately equal to the number on A divided by the number on S, and set it up on S in the rth position so that now S contains

$$2(10^{m-1}n_m + \ldots + 10^r n_{r+1}) + 10^{r-1}n_r.$$

Subtract n_r times, then add n_r to S in the rth position.

2.3 Types of Errors in Computations. Error Analysis

The numerical representation of a quantity x is usually subject to errors of various types.

(1) *Inherent errors.* When the quantity x is directly determined by physical measurements, then the error depends upon the precision of the measuring instruments.

(2) *Rounding errors.* These are due to the limitations of the calculating aids: tables, slide rule, desk machine or digital computer. Numbers have to be rounded off as they cannot exceed the capacity of the calculating aid.

(3) *Truncation errors.* They arise when x is evaluated in the form of an infinite series which is truncated at a certain stage when the remainder is presumed to be sufficiently small. The study of this type of error is sometimes called the convergence problem.

(4) *Blunders.* These are due usually to human errors during transcription or data preparation, but they may be due to some machine malfunctioning. In §2.5 we discuss possible checks to guard against them.

(5) Finally and more usually x may be obtained as the result of a calculation involving several steps each of which is subject to errors of one or more of the above types; then one has to study the propagation of these errors over a possibly large number of steps. This is sometimes called the stability problem.

Careful error analysis of each numerical procedure is beyond the scope of this book, but we give occasionally some indication of the appropriate error analysis. In the rest of this chapter we study errors in general.

The general problem may be formulated as follows. An unknown x is found in terms of known quantities

$$x_1, x_2, x_3, \ldots$$

$$x = f(x_1, x_2, \ldots) \qquad (2.3.1)$$

In this relation x is the result, f denotes a computing procedure, and $(x_1, x_2, \ldots)$ is the data; more precisely x is the exact answer which would be obtained by applying the exact numerical procedure f on the data $(x_1, x_2, \ldots)$. In fact the computed answer is x' where

$$x' = x + \delta x = f'(x_1, x_2, \ldots)$$

Here f' denotes the numerical procedure subject to various types of errors and thus is different from the theoretical f. The basic problem of error analysis is to find a bound, preferably a least upper bound, to the numerical value of δx, i.e. a bound to $|x' - x|$. This is called the forward or direct approach: given the accuracy in the arithmetic operations and the numerical representation of the data $(x_1, x_2, \ldots)$ we seek to find the accuracy in x; alternatively we seek to determine the necessary accuracy in the arithmetic and data in order to attain a prescribed accuracy in x. Often this problem is too difficult to answer directly; instead we seek to estimate $(x_1', x_2', \ldots)$ such that

$$x' = f(x_1', x_2', \ldots) \qquad (2.3.2)$$

i.e. x' is the result of carrying out the exact procedure f on the data $(x_1', x_2', \ldots)$. More precisely, if

$$x_i' = x_i + \delta x_i$$

then we seek a bound, or a least upper bound, to the numerical value of each δx_i. This is called backward or indirect error analysis. It is sometimes more amenable to analysis because it gives an estimate of the error in the computed answer x' by subtracting (2.3.1) from (2.3.2):

$$\delta x \sim \frac{\partial f}{\partial x_1} \delta x_1 + \frac{\partial f}{\partial x_2} \delta x_2 + \ldots$$

But usually there is no simple explicit formula f and the problem of estimating δx becomes difficult. Even for elementary problems there is no easy general rule. Often it helps to repeat the computation with two separate sets of data $(x_1, x_2, \ldots)$ and $(x_1', x_2', \ldots)$ where each disturbance $|x_i - x_i'|$ is equal to the maximum tolerable "noise". If the two answers differ substantially then this should be taken as a danger signal and the problem needs more attention. If on the other hand the two answers are each given to, say, 7 significant figures and they agree only, when rounded, to 5 significant figures then the answer is usually (but not necessarily) correct to only 5 significant figures.

2.4 Rounding Errors and Accuracy

On desk machines and digital computers, quantities cannot exceed a certain number of significant figures determined by the capacity of the machine. Suppose, for example, that the capacity of the machine is 4 decimal digits, then a number containing not more than 4 significant figures is said to be of single length. If it contains 8 significant figures it is said to be a double length number. To round off a double length number to a single length number on an automatic computer the usual procedure is to add half a unit of the least significant figure of the more significant half and chop off the resulting least significant half. For example,

$$4635 \times 3467 = 16069545$$

$$16069545 + 5000 = 16074545 \sim 16070000.$$

On some older digital computers, where the accumulator is single length with no guard figure, this form of rounding off is not possible because only the more significant half of the product is formed thus

$$4635 \times 3467 \text{ is given as } 1606 \times 10^4$$

Then a suitable rounding off procedure is to force the least significant digit to be even by adding 1 if necessary.

Thus

$$4635 \times 3467 (= 16069545) \text{ is given as } 1606 \times 10^4$$

and

$$4635 \times 5099 (= 23633865) \text{ is given as } 2364 \times 10^4$$

The general effect is to underestimate some products and overestimate others so that the overall effect of round off errors is

2

not systematically biased and cancels out as much as possible. An equally good procedure is to force the last significant digit to be odd.

Generally, we only know an approximation x' to any number x which may occur in a calculation. The error is

$$\delta x = x' - x$$

and $x - x'$ is the correction. Usually we do not know δx, but we may know that it does not exceed numerically an upper bound Δx which measures the accuracy or precision, so that

$$|\delta x| < \Delta x,$$
$$x - \Delta x < x' < x + \Delta x.$$

Thus the actual error is δx, and Δx is the error bound; sometimes one may also, less precisely, call Δx the error.

Example. If the number x is rounded to n decimal places then

$$\Delta x = \tfrac{1}{2} 10^{-n}.$$

If x is rounded to n significant figures then

$$\Delta x = \tfrac{1}{2} \times \text{least significant unit.}$$

If $x = 7 \cdot 32$ correct to 2 decimal places then $\Delta x = 0 \cdot 005$; if $x = 73200$ correct to 3 significant figures, then $\Delta x = 50$.

Sometimes δx is also called the absolute error and Δx is said to measure the absolute accuracy, where

$$\frac{\delta x}{x} \sim \frac{\delta x}{x'} \text{ is the relative error}$$

and

$$\frac{\Delta x}{|x|} \sim \frac{\Delta x}{|x'|} \text{ measures the relative accuracy.}$$

Thus if $x = 0 \cdot 52$ is given rounded to 2 decimal places then the absolute accuracy is $\Delta x = 0 \cdot 005$ and the relative accuracy is $0 \cdot 005/0 \cdot 52 \sim 1\%$.

When multiplying (or dividing) two or more numbers, the relative accuracy of the product is the sum of the relative accuracies of the factors. Thus, if x and y are positive, then

$$\Delta(xy) = (x + \Delta x)(y + \Delta y) - xy \sim x\Delta y + y\Delta x.$$

Hence

$$\frac{\Delta(xy)}{xy} \sim \frac{\Delta x}{x} + \frac{\Delta y}{y}.$$

Similarly,

$$\Delta\left(\frac{x}{y}\right) = \frac{x + \Delta x}{y - \Delta y} - \frac{x}{y} \sim \frac{y\Delta x + x\Delta y}{y^2}$$

(the signs are chosen to give the maximum possible error).
Hence

$$\Delta\left(\frac{x}{y}\right) \bigg/ \left(\frac{x}{y}\right) \sim \frac{\Delta x}{x} + \frac{\Delta y}{y}.$$

In addition (or subtraction) the absolute errors are added:

$$\Delta(x + y) = \Delta x + \Delta y.$$

But it is not possible to estimate the relative accuracy of the sum in terms of the relative accuracies of the numbers added only. Thus if $x = 7 \cdot 32$, $y = -4 \cdot 28$ are each correct to 2 decimal places, then $x + y = 3 \cdot 04$ and

$$\frac{\Delta x}{x} = \frac{0 \cdot 005}{7 \cdot 32} \sim 0 \cdot 07\%, \qquad \frac{\Delta y}{y} = \frac{0 \cdot 005}{4 \cdot 28} \sim 0 \cdot 12\%.$$

Then the relative error of the sum is

$$\frac{\Delta(x + y)}{x + y} = \frac{0 \cdot 01}{3 \cdot 04} \sim 0 \cdot 33\%.$$

An important source of large relative errors occurs when x and y are of opposite signs and numerically nearly equal. Thus if $x = 5 \cdot 37$ and $y = -5 \cdot 32$, each correct to 2 decimal places, then although the relative accuracy in x and y is of the order $0 \cdot 1\%$, the relative accuracy in $x + y$ is of the order 20%. *Ill-conditioned* or *unstable* problems often involve this type of cancellation. A simple example of an ill-conditioned problem is the solution of two linear simultaneous equations:

$$a_1x + b_1y = c_1$$

$$a_2x + b_2y = c_2$$

which represent nearly coincident straight lines. A small change in a coefficient causes a large change in the answer. This is because:

$$x = \frac{c_1b_2 - c_2b_1}{a_1b_2 - a_2b_1}$$

$$y = \frac{a_1c_2 - a_2c_1}{a_1b_2 - a_2b_1}$$

and $a_1b_2 \sim a_2b_1$, so that the denominator is very small as compared with a_1b_2 or a_2b_1.

Ill-conditioned problems are usually much more involved and difficult to detect. But this elementary example exhibits the main property of such problems: a small change in the data of the problem causes a large change in the answer. Sometimes this trouble arises because the problem is badly formulated and a different formulation may give better conditioned and more stable equations or expressions; Exercises 8 and 9 at the end of this chapter are simple examples of this. But sometimes the trouble is more intrinsic because of the sensitive nature of the problem and one may have to use double length or multiple length arithmetic.

We give a few more examples of loss of accuracy which occur frequently in computations. Suppose we wish to compute $(a - b)^{\frac{1}{2}}$ when $a = 0 \cdot 5394$, $b = 0 \cdot 5311$ rounded to 4 decimal places. Working in fixed point 4-figure arithmetic, we have:

$$a - b = 0 \cdot 0083 \qquad (a - b)^{\frac{1}{2}} = 0 \cdot 0911$$

As loss of accuracy occurs in subtraction, the final answer is only correct to 3 decimal places, so the last digit is misleading and spurious. Similarly

$$(a - b)^{\frac{1}{4}} = 0 \cdot 2025$$

has two spurious digits; only the first two digits are significant.

The same phenomenon occurs, though not so explicitly, in the following problem: to find sin x, given that cos $x = 0 \cdot 9957$ correct to 4 decimal places. We have:

$$\sin^2 x = 1 - \cos^2 x = 0 \cdot 0086$$

and

$$\sin x = 0 \cdot 0927$$

which is correct to only 3 decimal places.

Next we consider a problem where loss of accuracy can be avoided if we take the necessary precautions. The simultaneous equations:

$$x^2 - y^2 = 2a \qquad (2.4.1)$$

$$xy = 2b \qquad (2.4.2)$$

occur frequently, for example when finding the square root of a complex number, or when finding sin θ and cos θ when we know tan 2θ. If we eliminate x we get:

$$y^4 + 2ay^2 - b^2 = 0$$
$$y^2 = - a + (a^2 + b^2)^{\frac{1}{2}} \qquad (2.4.3)$$

If we eliminate y we get:

$$x^4 - 2ax^2 - b^2 = 0$$
$$x^2 = a + (a^2 + b^2)^{\frac{1}{2}} \qquad (2.4.4)$$

If a is positive we use (2.4.4) to calculate x and then use (2.4.2) to calculate y. If a is negative we use (2.4.3) to calculate y then use (2.4.2) to calculate x. This avoids cancellation which can be considerable if b is numerically much smaller than a. For example, if $a = - 1$ and $b = 0 \cdot 01$, equation (2.4.4) gives

$$x^2 = - 1 + (1 \cdot 0001)^{\frac{1}{2}};$$

this will have considerable cancellation; in fact even if we use 5-figure fixed point arithmetic, it will give $x = 0$. But if we use equation (2.4.3) to find y first we get $y = 2 \cdot 0000$ and $x = 0 \cdot 0100$ which will satisfy (2.4.1) with an error equal to $0 \cdot 0001$.

2.5 Checks

Checks are necessary in all numerical work, whether on desk machines or digital computers. One has to check the arithmetic, the accuracy, and the method.

To ensure that no arithmetic mistakes are committed (by being very careful) is laborious and ineffective. It is better, whenever possible, to devise a check; then one can carry out the arithmetic faster and repeat the work and check if the first check does not succeed. A solution which requires prolonged arithmetical work should be broken into intermediate stages and each stage checked before proceeding to the next one. A method which does not lend itself to this sort of check is a bad method, especially if it is carried out on desk machines. Arithmetic checks are not so necessary on digital computers as such mistakes occur rarely and when they do they are caused by malfunctioning of some components; usually, then, a computer will automatically stop and give warning.

Example 1. To add 20 numbers $a_1, a_2, \ldots, a_{20}$, they are arranged in 5 rows and 4 columns. First we find the sum of each row: $r_1, r_2, \ldots, r_5$, and then the sum of each column: c_1, c_2, c_3, c_4. The required sum is

$$r_1 + r_2 + r_3 + r_4 + r_5 = c_1 + c_2 + c_3 + c_4.$$

We check that the two sums are the same. This method amounts to doing the same arithmetic twice but in a different order.

Example 2. Given the vectors $\mathbf{a}, \mathbf{b}_1, \mathbf{b}_2, \ldots, \mathbf{b}_n$, where each vector is an array of numbers, it is required to find each of the scalar products $\mathbf{a} \cdot \mathbf{b}_1, \mathbf{a} \cdot \mathbf{b}_2, \ldots, \mathbf{a} \cdot \mathbf{b}_n$. These are calculated in the usual manner. Then we compute the sum vector

$$\mathbf{s} = \mathbf{b}_1 + \mathbf{b}_2 + \ldots + \mathbf{b}_n$$

and the scalar product $\mathbf{a} \cdot \mathbf{s}$, and check that

$$\mathbf{a} \cdot \mathbf{s} = \mathbf{a} \cdot \mathbf{b}_1 + \mathbf{a} \cdot \mathbf{b}_2 + \ldots + \mathbf{a} \cdot \mathbf{b}_n.$$

In this example the extra arithmetic involved in checking is a small fraction of the total work.

Solutions based on converging iterative methods do not need arithmetic checks, as big mistakes are quickly spotted and small arithmetic mistakes do not alter the final answer although the solution may require more iterations.

To check the accuracy of an answer, say $x = 3 \cdot 72$ claimed to be correct to 2 decimal places, means to check that the true answer lies between $3 \cdot 715$ and $3 \cdot 725$. This may be possible in a simple case such as when x is the zero of a simple polynomial; it is sufficient to evaluate the polynomial for $x = 3 \cdot 715$ and $x = 3 \cdot 725$ and check that the polynomial changes sign, but even in such a simple case we have to ensure that the rounding errors in evaluating the polynomial do not alter the sign of the final value. More often this back substitution is impossible or unreliable. To check accuracy on desk machines it is advisable to carry out the arithmetic with one or two extra guard figures whenever possible. On digital computers one may have to use double length arithmetic. In problems of integration one can repeat the integration with half the interval. In many situations it is not easy to check the accuracy and the check merges into the problem of error analysis.

The method may have to be checked also. A method which works in one case may not work in another. There is seldom a method which does not fail for some cases. One may solve the problem by a second method to check the arithmetic, accuracy and the first method.

On digital computers checks are particularly necessary when one is often tempted to use existing library programs which may not be suitable, and where one usually has no information on intermediate internal results which can give a lot of useful guidance to a human when operating a desk machine. Also a program may contain a subtle logical flaw which escapes detection and gives correct results for many cases. On the other hand checks are correspondingly easier to apply in digital computers.

EXERCISES

1. Find correct to 6 significant figures the reciprocal of the following numbers (a) by division, and (b) by building up 1 from multiples of the given number: $1 \cdot 53946$, $0 \cdot 0783267$, $592 \cdot 614$.

2. Evaluate the polynomial

$$3 \cdot 742x^4 - 2 \cdot 579x^3 + 8 \cdot 942x^2 + 4 \cdot 701x + 1 \cdot 387$$

for $x = 1 \cdot 348$ and $x = -1 \cdot 348$. Carry out the arithmetic to 4 decimal places.

3. Find $\sin x$ and $\cos x$ when $x = 1 \cdot 2$ radians correct to 6 decimal places using the series

$$\sin x = x - x^3/3! + x^5/5! - \ldots$$
$$\cos x = 1 - x^2/2! + x^4/4! - \ldots$$

   [*Suggestion:* If $u = 1/x^2$ then

$$\sin x = x^{11}(u^5 - u^4/3! + u^3/5! - u^2/7! + u/9! - 1/11!).]$$

   [*Ans.* $0 \cdot 932039$, $0 \cdot 362358$.]

4. Find correct to 6 significant figures the square roots of the following numbers using desk machines and check by squaring:

$$37459, \quad 78 \cdot 382, \quad 7 \cdot 8382, \quad 0 \cdot 000351, \quad 10 \cdot 2$$

5. Round off to 3 significant figures the following

$$0 \cdot 006435, \quad 0 \cdot 006425, \quad 0 \cdot 7289, \quad 99 \cdot 99$$

   [When the digit discarded is exactly 5, a useful rule is to force the last figure of the rounded number to be even: thus $0 \cdot 006435$ is rounded to $0 \cdot 00644$ and $0 \cdot 006425$ is rounded to $0 \cdot 00642$. This gives more accuracy than the rule which adds 1 whenever the discarded figure is 5 or more. Why?]

6. A number is rounded off to n significant figures. Show that the maximum relative round-off error is $5/10^n$. Verify this result for the following numbers rounded to 3 significant figures: $0 \cdot 0965$, 105000.

7. Evaluate the following expressions where each number is rounded off to 2 decimal places. Give the answers correct to the maximum number of significant figures, discarding uncertain figures:

$$\frac{3 \cdot 72 \times 2 \cdot 48}{4 \cdot 79}, \quad \frac{3 \cdot 72 \times 2 \cdot 48}{0 \cdot 48}, \quad \frac{3 \cdot 72 \times 2 \cdot 48}{0 \cdot 05}$$

   [*Hint.* First method:

$$\frac{3 \cdot 725 \times 2 \cdot 485}{4 \cdot 785} = 1 \cdot 935$$

$$\frac{3 \cdot 715 \times 2 \cdot 475}{4 \cdot 795} = 1 \cdot 918$$

The answer lies between these two limits so that the result is $1 \cdot 9$.
Second method:

$$\frac{3 \cdot 72 \times 2 \cdot 48}{4 \cdot 79} = 1 \cdot 926$$

The relative error is

$$\frac{0 \cdot 005}{3 \cdot 27} + \frac{0 \cdot 005}{2 \cdot 48} + \frac{0 \cdot 005}{4 \cdot 75} = 0 \cdot 0044$$

The absolute error is $1 \cdot 926 \times 0 \cdot 0044 = 0 \cdot 008$, the answer lies between $1 \cdot 926 + 0 \cdot 008 = 1 \cdot 934$ and $1 \cdot 926 - 0 \cdot 008 = 1 \cdot 918$.]

8. Find the smaller root of the equation

$$x^2 - 4x + 0 \cdot 00643 = 0$$

correct to 3 significant figures:
   (a) By using the formula for quadratic equations.
   (b) By ignoring the term in x^2.
   [*Ans.* $0 \cdot 00161$. This is a case where the formula is an ill-conditioned expression, and the square root has to be worked out to 6 significant figures to get the answer correct to 3 significant figures.]

9. Prove that the two expressions

$$\frac{x_n x_{n+2} - x_{n+1}^2}{x_{n+2} - 2x_{n+1} + x_n}, \qquad x_{n+2} - \frac{(x_{n+2} - x_{n+1})^2}{x_{n+2} - 2x_{n+1} + x_n}$$

are identical. They are used to find the limit of a sequence of numbers x_n, x_{n+1}, x_{n+2}, ... Evaluate the two expressions when $x_n = 3 \cdot 572$, $x_{n+1} = 3 \cdot 621$, $x_{n+2} = 3 \cdot 649$, using four figure tables.
   [*Ans.* The first expression $= 3 \cdot 714$, the second expression $= 3 \cdot 686$. If we use a desk machine or 8 figure table then both expressions give the better answer $3 \cdot 686$. This is a case where the first expression is ill-conditioned but by rearranging the expression we can reduce it to a better conditioned expression which gives more accuracy for the same number of figures in the arithmetic.]

10. If $x = 25 \cdot 3°$ correct to the first decimal place, find $\sin x$ and $\cos x$ rounded off to the maximum number of significant figures, discarding all the uncertain figures. Find similarly e^x, $\sinh x$, and $\cosh x$ when $x = 1 \cdot 52$ is correct to 2 decimal places.
   [*Hint.* $\sin 25 \cdot 25° = 0 \cdot 4265$, $\sin 25 \cdot 35° = 0 \cdot 4281$, thus the answer is $0 \cdot 43$. Alternatively $\sin 25 \cdot 3° = 0 \cdot 4273$, $\delta x = 0 \cdot 05° = 0 \cdot 00087$ radians. $\delta (\sin x) = (\cos x) \delta x = 0 \cdot 0008$, then $\sin x = 0 \cdot 4273 \pm 0 \cdot 0008$.]

CHAPTER 3

Finite-Difference Methods

3.1 Difference Tables

Given a function y of x (such as $y = e^x$ in Table 1 or $y = \sin x$ in Table 2) a difference table can be constructed if y is given for equally spaced values of x. This constant interval or step length in x is often denoted by h. Thus $h = 0 \cdot 1$ in Table 1, and $h = 5$ in Table 2. The difference table is made up of columns; the first column lists the values of x, the second column lists the corresponding values of y. The next column is the column of first differences denoted by Δy, the entries in this column being derived from the y column by differencing it; each entry in the Δy column is equal to the difference of the two entries on its left (the *lower entry* minus the *upper entry*) and is usually placed in a level midway between these two entries. The column of second differences, denoted by $\Delta^2 y$, is similarly obtained from the Δy column, and so on for higher differences. All the entries are given to the same number of decimal places and a useful convention is to drop the decimal point and any zeros on the left. Thus, to 6 decimal places, the entry -650 means $-0 \cdot 000650$, and -65 means $-0 \cdot 000065$. Another useful practice is to carry on one or two guard figures and use a comma as a subsidiary decimal point. Thus if, in Table 1, we need the final answers of a computation to be correct to 4 decimal places, we tabulate to 6 decimal places and the entries along $x = 0 \cdot 6$, say, would be in the form:

$$1 \cdot 8221{,}19 \qquad 182{,}36 \qquad 1{,}83.$$

The beginner should study Tables 1 and 2 and note the gradual variation of consecutive entries in each column and the way entries get smaller in absolute values for higher differences up to a

TABLE 1

$y = e^x$

x	y	Δy	$\Delta^2 y$	$\Delta^3 y$	$\Delta^4 y$	$\Delta^5 y$	$\Delta^6 y$	$\Delta^7 y$
0·0	1·000000							
		105171						
0·1	1·105171		11061					
		116232		1163				
0·2	1·221403		12224		123			
		128456		1286		11		
0·3	1·349859		13510		134		7	
		141966		1420		18		−15
0·4	1·491825		14930		152		−8	
		156896		1572		10		20
0·5	1·648721		16502		162		12	
		173398		1734		22		−16
0·6	1·822119		18236		184		−4	
		191634		1918		18		7
0·7	2·013753		20154		202		3	
		211788		2120		21		−2
0·8	2·225541		22274		223		1	
		234062		2343		22		6
0·9	2·459603		24617		245		7	
		258679		2588		29		−10
1·0	2·718282		27205		274		−3	
		285884		2862		26		9
1·1	3·004166		30067		300		6	
		315951		3162		32		−1
1·2	3·320117		33229		332		5	
		349180		3494		37		−6
1·3	3·669297		36723		369		−1	
		385903		3863		36		10
1·4	4·055200		40586		405		9	
		426489		4268		45		−9
1·5	4·481689		44854		450		0	
		471343		4718		45		9
1·6	4·953032		49572		495		9	
		520915		5213		54		
1·7	5·473947		54785		549			
		575700		5762				
1·8	6·049647		60547					
		636247						
1·9	6·685894							

TABLE 2

$x°$	y	Δy	$\Delta^2 y$	$\Delta^3 y$	$\Delta^4 y$	$\Delta^5 y$	$\Delta^6 y$	$\Delta^7 y$
0	0·000000							
		87156						
5	0·087156		−664					
		86492		−657				
10	0·173648		−1321		8			
		85171		−649		8		
15	0·258819		−1970		16		−4	
		83201		−633		4		3
20	0·342020		−2603		20		−1	
		80598		−613		3		7
25	0·422618		−3216		23		6	
		77382		−590		9		−18
30	0·500000		−3806		32		−12	
		73576		−558		−3		26
35	0·573576		−4364		29		14	
		69212		−529		11		−23
40	0·642788		−4893		40		−9	
		64319		−489		2		6
45	0·707107		−5382		42		−3	
		58937		−447		−1		14
50	0·766044		−5829		41		11	
		53108		−406		10		−25
55	0·819152		−6235		51		−14	
		46873		−355		−4		25
60	0·866025		−6590		47		11	
		40283		−308		7		−17
65	0·906308		−6898		54		−6	
		33385		−254		1		5
70	0·939693		−7152		55		−1	
		26233		−199		0		3
75	0·965926		−7351		55		2	
		18882		−144		2		
80	0·984808		−7495		57			
		11387		−87				
85	0·996195		−7582					
		3805						
90	1·000000							

certain order, beyond which they become erratic and begin to increase again. If the interval of tabulation is small enough, the gradual variation and initial decrease in magnitude are characteristic of well-behaved functions and when they do not occur it is usually because an arithmetic blunder has taken place somewhere. In the next section we show how to detect, locate and correct such blunders and estimate bounds to permissible errors due to round off.

The consecutive entries in the x and y columns are labelled by suffices from some convenient origin:

$$\ldots, y_{-2}, y_{-1}, y_0, y_1, y_2, \ldots$$

Then
$$\Delta y_0 = y_1 - y_0,$$
$$\Delta y_1 = y_2 - y_1,$$

and in general
$$\Delta y_r = y_{r+1} - y_r \qquad (3.1.1)$$

so that
$$y_{r+1} = y_r + \Delta y_r. \qquad (3.1.2)$$

Similar rules apply for higher differences

$$\Delta^2 y_r = \Delta y_{r+1} - \Delta y_r,$$
$$\Delta^3 y_r = \Delta^2 y_{r+1} - \Delta^2 y_r,$$

and in general we have

$$\Delta^m y_r = \Delta^{m-1} y_{r+1} - \Delta^{m-1} y_r, \qquad (3.1.1')$$

$$\Delta^m y_{r+1} = \Delta^m y_r + \Delta^{m+1} y_r. \qquad (3.1.2')$$

In these equations the indices 2, 3, m are not powers, they denote higher differences as defined recursively by (3.1.1'). The subscript r is not uniquely determined and depends upon which value of x is taken as the origin x_0. Thus if in Table 1 $x = 0 \cdot 6$ is taken as x_0 then $0 \cdot 9$ is x_3 and the corresponding value of $y = 2 \cdot 459603$ is y_3. But if $x = 0 \cdot 8$ is taken as x_0 then $0 \cdot 9$ is x_1 and $2 \cdot 459603$ is y_1.

Table 3 shows the arrangement of the subscripts round about y_0. Note that the subscript is constant along a downward sloping diagonal so that y_0, Δy_0, $\Delta^2 y_0$, $\Delta^3 y_0$, $\ldots$, are in line diagonally.

TABLE 3. FORWARD DIFFERENCES LAYOUT

x_{-2}	y_{-2}				
		Δy_{-2}			
x_{-1}	y_{-1}		$\Delta^2 y_{-2}$		
		Δy_{-1}		$\Delta^3 y_{-2}$	
x_0	y_0		$\Delta^2 y_{-1}$		$\Delta^4 y_{-2}$
		Δy_0		$\Delta^3 y_{-1}$	
x_1	y_1		$\Delta^2 y_0$		
		Δy_1			
x_2	y_2				

There are many more useful relations, which are a little more complicated than (3.1.1) and (3.1.2). For example:

$$\Delta^2 y_0 = \Delta y_1 - \Delta y_0 = (y_2 - y_1) - (y_1 - y_0),$$

so that

$$\Delta^2 y_0 = y_2 - 2y_1 + y_0. \tag{3.1.3}$$

This and many other more involved formulae can be derived more easily by the use of the concept of the operators E and Δ. The operator E is defined by the relation

$$E y_r = y_{r+1}.$$

It is the operator which advances the suffix by 1. The relation

$$\Delta y_r = y_{r+1} - y_r$$

can be put in the form

$$\Delta y_r = E y_r - y_r = (E - 1) y_r.$$

Hence we have the relation

$$\Delta = E - 1, \tag{3.1.4}$$

and

$$E = 1 + \Delta. \tag{3.1.5}$$

The beginner is warned not to take an expression such as $(E - 1) y_r$ literally as the product of two numbers $E - 1$ and y_r. The expression $(E - 1) y_r$ is symbolic for $E y_r - y_r = y_{r+1} - y_r$. The expression $E^2 y_r$ stands for y_{r+2} and similarly for higher powers. Note also that the operator 1, in the expression $(E-1) y_r$, means take y_r and is not a unit.

By raising the operators in (3.1.4) to various powers we get

$$\Delta^2 y_0 = (E - 1)^2 y_0 = (E^2 - 2E + 1) y_0 = y_2 - 2y_1 + y_0.$$

Similarly

$$\Delta^3 y_0 = y_3 - 3y_2 + 3y_1 - y_0.$$

By raising the operators in (3.1.5) to various powers we get

$$y_2 = E^2 y_0 = (1 + \Delta)^2 y_0 = y_0 + 2\Delta y_0 + \Delta^2 y_0.$$

Similarly

$$y_3 = E^3 y_0 = y_0 + 3\Delta y_0 + 3\Delta^2 y_0 + \Delta^3 y_0,$$

and generally

$$y_p = E^p y_0 = (1 + \Delta)^p y_0$$

Hence

$$y_p = y_0 + \binom{p}{1} \Delta y_0 + \binom{p}{2} \Delta^2 y_0 + \binom{p}{3} \Delta^3 y_0 + \dots \qquad (3.1.6)$$

This is an important relation, it is called the *Gregory–Newton formula for forward interpolation*. In this equation the expression $\binom{p}{r}$ is the coefficient of x^r in the binomial expansion of $(1 + x)^p$.

Sometimes the binomial coefficient $\binom{p}{r}$ is denoted by ^pC_r. Thus:

$$\binom{p}{1} = p$$

$$\binom{p}{2} = p(p - 1)/2!$$

$$\binom{p}{3} = p(p - 1)(p - 2)/3! = (p^3 - 3p^2 + 2p)/6$$

For the moment we assume that p is a positive integer so that the right-hand side of (3.1.6) terminates after $(p + 1)$ terms, as $\binom{p}{r} = 0$ when $r > p$.

We note that (3.1.6) has been *derived* not *proved*; the derivation was by purely formal algebraic manipulation of the operators E and Δ; a more careful justification is given in §3.3. We also note that expressions in E and Δ operate on y and are not commutative with y.

Exercise. Verify (3.1.6) for various integral values of p taking as y_0 various entries in the y column of Tables 1 and 2.

Note. Whether using desk machines or mental arithmetic, mistakes occur frequently in the construction of difference tables. As each difference column is worked out the following check is often useful and is carried out before proceeding to the next column, or when locating an error that has been detected.

$$\Delta y_1 + \Delta y_2 + \ldots + \Delta y_n = (y_2 - y_1) + (y_3 - y_2) + \ldots + (y_{n+1} - y_n)$$
$$= y_{n+1} - y_1.$$

The same check holds for differences of higher order: the sum of each column = bottom entry − top entry in previous column. Apply this check to the difference columns of Tables 1 and 2.

3.2 Differencing Polynomials: Errors in Difference Tables

If y is a polynomial of degree n in x, then Δy is a polynomial of degree $n - 1$ in x. For example,

$$\Delta x^3 = (x + h)^3 - x^3 = 3hx^2 + 3h^2x + h^3$$

is a polynomial of degree 2. It follows that the nth difference of a polynomial of degree n is constant and the $(n + 1)$th difference is 0. This property can be used to tabulate a polynomial at equal intervals.

Suppose, for example, that it is required to tabulate the polynomial

$$y = x^3 - 3x + 1 \text{ for } x = 1(0\cdot1)\,2.$$

(This notation, which is frequently used, means for values of x from 1 to 2 at intervals of $0\cdot1$.) Only the first 4 values of y need to be worked out, though as an overall check it would be wise to find a 5th value and then check that the two third differences are the same.

Starting then with the entries in Table 4 above the line, the remaining entries are rapidly worked out by simple additions using the fact that all the third differences are 6 (i.e. 0·006). Thus:

$$6 + 78 = 84,$$

$$84 + 247 = 331,$$

$$331 - 456 = -125,$$

i.e. $y = -0·125$ when $x = 1·5$. Similarly

$$6 + 84 = 90$$

$$90 + 331 = 421$$

$$421 - 125 = 296$$

giving the next value of y and so on.

TABLE 4

x	y		2	3
	$y = x^3 - 3x + 1$			
1·0	−1·000			
		31		
1·1	−0·969		66	
		97		6
1·2	−0·872		72	
		169		6
1·3	−0·703		78	
		247		6
1·4	−0·456		84	
		331		6
1·5	−0·125		90	
		421		6
1·6	0·296		96	
		517		6
1·7	0·813		102	
		619		6
1·8	1·432		108	
		727		6
1·9	2·159		114	
		841		
2·0	3·000			

Exercise. If $\quad y = a_0x^n + a_1x^{n-1} + a_2x^{n-2} + \ldots + a_n$

is a polynomial of degree n tabulated exactly at intervals of constant length h show that
$$\Delta^n y = a_0 h^n n!$$
Verify this result for Table 4.

Difference tables can be used to check a function tabulated at equal intervals. Sometime this can be done by sight. Thus if the successive entries of a table are

$$0\cdot259 \quad 0\cdot342 \quad 0\cdot423 \quad 0\cdot500 \quad 0\cdot547 \quad 0\cdot643 \quad 0\cdot707$$

We notice that the successive differences are

$$83 \quad\quad 81 \quad\quad 77 \quad\quad 47 \quad\quad 96 \quad\quad 64$$

It is obvious that there must be an error in the entry $0\cdot547$. If an error in the table is small it is not detected from the first differences, but it may be detected from higher differences.

A small error in the y-column causes larger errors in the columns of higher differences. To see the result of replacing y_r by $y_r + e$ where e is a small error, we consider the following difference table:

y	Δ	Δ^2	Δ^3	Δ^4
0				
	0			
0		0		
	0		0	
0		0		e
	0		e	
0		e		$-4e$
	e		$-3e$	
e		$-2e$		$6e$
	$-e$		$3e$	
0		e		$-4e$
	0		$-e$	
0		0		e
	0		0	
0		0		
	0			
0				

The entries in the column of the nth differences are multiples of e, the coefficients being the binomial coefficients in the expansion of $(a - b)^n$. Thus 1, -4, 6, -4, 1 are the coefficients of the expansion of $(a - b)^4$. This indicates the way an error propagates to higher differences. It causes a rapidly increasing oscillatory disturbance in an otherwise smooth trend. This property can be used to detect, locate and correct an error in the y-column or in a difference column. Thus if the entries in the column of the 4th differences are

$$0 \quad 0 \quad 10 \quad -40 \quad 60 \quad -40 \quad 10 \quad 0 \quad 0$$

we recognize that the entry in the y-column in the same level as 60 should be reduced by 10. A frequent form of error in the y-column is due to the transposition of adjacent digits, the error is then a multiple of 9. But usually the effect of an error does not take quite the binomial pattern because the corrected entries are not precisely 0 except for exact polynomials and because round-off errors may distort the binomial pattern. The effect of round-off errors in the given tabulated values in the y-column may be relatively large if y is given to a small number of significant figures; otherwise the effect of round-off errors in consecutive values of y is irregular. A reasonable estimate of this effect in the nth difference is m units of the least significant digit, where m is the largest coefficient in the expansion of $(a + b)^n$.

n	4	5	6	7	8
m	6	10	20	35	70.

A much larger disturbance indicates the presence of an error other than round-off.

Finally, differences may not converge satisfactorily because the interval h is too large, or because the function is irregular, or has a singularity in or near the investigated range.

3.3 Interpolation

We recall the Gregory–Newton formula (3.1.6):

$$y_p = y_0 + \binom{p}{1} \Delta y_0 + \binom{p}{2} \Delta^2 y_0 + \binom{p}{3} \Delta^3 y_0 + \ldots$$

The right-hand side will terminate if y is a polynomial because all differences above the order of the polynomial are 0 even though p is not a positive integer. Thus to find y when $x = 1 \cdot 24$ in Table 4, we take $x_0 = 1 \cdot 2$, $x_p = 1 \cdot 24$. This gives $p = 0 \cdot 4$ from the relation

$$x_p = x_0 + hp.$$

Here $h = 0 \cdot 1$. Substituting these values in the formula we get:

$$
\begin{aligned}
y_p= \quad\quad & -0 \cdot 872 = -0 \cdot 872 \quad = -0 \cdot 813376 \\
& +0 \cdot 4 \times 0 \cdot 169 \quad +0 \cdot 0676 \\
& +\tfrac{1}{2} \times 0 \cdot 4 \times -0 \cdot 6 \times 0 \cdot 078 \quad -0 \cdot 00936 \\
& +\tfrac{1}{6} \times 0 \cdot 4 \times -0 \cdot 6 \times -1 \cdot 6 \times 0 \cdot 006 \quad +0 \cdot 000384.
\end{aligned}
$$

Check:

$$x^3 - 3x + 1 = 1 \cdot 24^3 - 3 \times 1 \cdot 24 + 1 = 0 \cdot 813376.$$

This process of computing y for an untabulated value of x is called *interpolation*. It is one of the basic problems in numerical analysis. Many functions are difficult and lengthy to compute directly for any value of x. Instead they are computed and tabulated only for some few values of x and for other values of x they are computed by interpolation. In this chapter we consider functions tabulated at equal intervals, and in Chapter 8 we consider unequal intervals. Interpolation is then a method of approximating a usually complicated function by the much simpler interpolation formula; from this one can derive formulae for approximate differentiation and integration. It is therefore worthwhile considering closely the validity of interpolation formulae as they can give incorrect results if they are misused.

The Gregory–Newton formula has been derived from the algebraic identity

$$\mathrm{E}^p = (1 + \Delta)^p = 1 + \binom{p}{1} \Delta + \binom{p}{2} \Delta^2 + \ldots .$$

But derivation is not a justification or proof; it is not possible to prove the Gregory–Newton formula without taking into account the nature of the function $y = f(x)$ which is being approximated and examining the remainder term when the infinite series in the formula is truncated. First we take the case when $f(x)$ is a polynomial of degree n. Then the right-hand side of (3.1.6) will terminate, and if we replace p by $(x - x_0)/h$, this right-hand side becomes a polynomial $g(x)$ of degree n which coincides with $f(x)$ at the $n + 1$ points x_0, x_1, ... , x_n when $p = 0, 1, 2, ... , n$ respectively. It follows from basic theorems on polynomials that $g(x)$ and $f(x)$ are identical and this establishes the Gregory–Newton formula for any other value of p.

Next we consider the case when y is not a polynomial and p is not an integer so that the series do not terminate. We may often approximate y by a polynomial for which the formula does work. The effect of truncating the series say at the 5th term is to approximate the given function by a polynomial of degree 5 which coincides with the given function at $x = x_0$, x_1, ... , x_5 so that the 6th and higher differences are zero for such a polynomial. The possibility of such an approximation may be investigated mathematically by considering the remainder term; this is done later on in §8.1 and in some of the proofs of the integration formulae in §3.9, but this can be difficult. In practice it suffices (a) to check that differences of higher order cease to contribute to the numerical result for the required number of significant figures, so that the series seem to converge; and (b) to check that, if the series is truncated at the term in $\Delta^n y$, then $\Delta^{n+1} y$ is negligible for several consecutive values of x.

Four or five terms should normally suffice, though more can be used for safety. Thus only 5 terms are taken if $\Delta^5 y_0$ is so small that the 6th term is negligible to the required accuracy; one can be confident that the result of the interpolation is accurate if $\Delta^5 y_r$ is of the same small order of magnitude as $\Delta^5 y_0$ for $r = -2$, $-1, 1, 2$. After a certain stage the terms in higher differences get larger owing to round-off errors. Danger often makes itself apparent if one does not get the required accuracy by that stage, and taking more terms in the formula will not improve the accuracy. To get more accuracy one has to tabulate the function at smaller intervals and to a larger number of decimal places.

Exercises. 1. Find from Table 1, correct to 6 decimal places, the values of e^x when $x = 0 \cdot 12$, $0 \cdot 75$, $1 \cdot 08$; in each case take as x_0 the nearest tabular point. Find from Table 2 $\sin x$ when $x = 6°$, $6° \, 20'$ and check from tables.

2. Find from Table 1, correct to 6 decimal places, e^x when $x = 0 \cdot 75$ by taking as x_0 each of the values $0 \cdot 4$, $0 \cdot 5$, $0 \cdot 6$, $0 \cdot 7$, $0 \cdot 8$, and compare convergence.

3.4 Inverse Interpolation

The problem of interpolation is to find y for a given value of x. The problem of inverse interpolation is to find x for a given value of y. This problem can arise in various contexts, for example in Table 1 to find x when $y = 2$ is to find $\ln 2$; in Table 4 to find x when $y = 0$ is to find the root of the equation $x^3 - 3x + 1 = 0$ which lies between 1 and 2. The problem would be difficult and indeterminate if y takes the same value for several neighbouring values of x; this happens if the function is rapidly oscillating. In the methods explained here it is assumed that y takes the given value for a unique x in the investigated neighbourhood.

In the formula

$$y_p = y_0 + \binom{p}{1} \Delta y_0 + \binom{p}{2} \Delta^2 y_0 + \binom{p}{3} \Delta^3 y_0 + \ldots$$

$$= y_0 + p\Delta y_0 + \tfrac{1}{2}(p^2 - p) \, \Delta^2 y_0 - $$
$$- \tfrac{1}{6}(p^3 - 3p^2 + 2p) \, \Delta^3 y_0 + \ldots$$

we are given y_p, y_0, Δy_0, $\Delta^2 y_0$, $\ldots$, and we require to find p which gives x from the formula

$$x = x_p = x_0 + ph.$$

The term $p\Delta y_0$ is the leading term containing the unknown p; the other terms containing p are of smaller order of magnitude which can be neglected to obtain a first approximation for p. The formula is put in the form

$$p = \{y_p - y_0 - \tfrac{1}{2}(p^2 - p) \, \Delta^2 y_0 - $$
$$- \tfrac{1}{6}(p^3 - 3p^2 + 2p) \, \Delta^3 y_0 + \ldots \}/\Delta y_0$$

or, briefly, $p = F(p)$. In general we obtain successively better approximations $p_0, p_1, p_2, \ldots$, to p by the formula

$$p_{r+1} = \{y_p - y_0 - \tfrac{1}{2}(p_r^2 - p_r)\,\Delta^2 y_0 - $$
$$- \tfrac{1}{6}(p_r^3 - 3p_r^2 + 2p_r)\,\Delta^3 y_0 + \ldots \}/\Delta y_0$$

or, briefly,

$$p_{r+1} = F(p_r).$$

Example. To find from Table 4 the root of $x^3 - 3x + 1 = 0$ which lies between 1 and 2. Here

$$x_0 = 1 \cdot 5, \qquad x_p = x_0 + hp = 1 \cdot 5 + 0 \cdot 1p.$$

The equation

$$y_p = y_0 + \binom{p}{1}\Delta y_0 + \binom{p}{2}\Delta^2 y_0 + \binom{p}{3}\Delta^3 y_0 + \ldots$$

becomes

$$0 = -0 \cdot 125 + 0 \cdot 421p + 0 \cdot 048p(p - 1) + $$
$$+ 0 \cdot 001p(p - 1)(p - 2).$$

Hence

$$0 \cdot 421 p_{r+1} = 0 \cdot 125 - 0 \cdot 048 p_r(p_r - 1) - $$
$$- 0 \cdot 001 p_r(p_r - 1)(p_r - 2),$$

or

$$p_{r+1} = 0 \cdot 296912 - 0 \cdot 114014 p_r(p_r - 1) - $$
$$- 0 \cdot 002375 p_r(p_r - 1)(p_r - 2).$$

We get

$$p_0 = 0$$
$$p_1 = 0 \cdot 296912$$
$$p_2 = 0 \cdot 319869$$
$$p_3 = 0 \cdot 320848$$
$$p_4 = 0 \cdot 320887$$

and $x = 1 \cdot 5 + 0 \cdot 1p = 1 \cdot 532089$ is correct to 6 decimal places.

An alternative method is to use subtabulation. Ignoring terms in $\Delta^2 y_0$ and higher differences we solve for p, then tabulate the function for a much smaller interval, say $0\cdot01h$, and repeat. In our example we have

$$y_p = y_0 + p\Delta y_0,$$
$$0 = -0\cdot125 + 0\cdot421p,$$

which gives $p = 0\cdot30$. Using the interpolation formula, or by direct evaluation, we compute y for $p = 0\cdot30$ and $p = 0\cdot31$, i.e. for $x = 1\cdot530$ and $x = 1\cdot531$. The corresponding values of y are $-0\cdot008423$ and $-0\cdot004396$ which gives

$$0 = -0\cdot008423 + 0\cdot004027p.$$

Hence

$$p = 2\cdot09 \text{ and } x = 1\cdot53209.$$

Finally we evaluate y for $x = 1\cdot53209$ and $1\cdot53210$; we get

$$y = 0\cdot000005 \text{ and } 0\cdot000044,$$
$$0 = 0\cdot000005 - 0\cdot000039p,$$
$$p = -0\cdot12 \text{ and } x = 1\cdot5320888.$$

The last answer is correct to 8 significant figures,

x_0	$1\cdot5$	$1\cdot530$	$1\cdot53209$
x_1	$1\cdot6$	$1\cdot531$	$1\cdot53210$
h	$0\cdot1$	$0\cdot001$	$0\cdot00001$
y_0	$-0\cdot125$	$-0\cdot008423$	$0\cdot000005$
y_1	$0\cdot296$	$-0\cdot004396$	$0\cdot000044$
Δy_0	$0\cdot421$	$0\cdot004027$	$0\cdot000039$
p	$0\cdot30$	$2\cdot09$	$-0\cdot12$
x	$1\cdot530$	$1\cdot53209$	$1\cdot5320888$

If two desk machines are available faster methods can be devised. We postpone this to §3.6 after obtaining Bessel's formula, which is the best one to use. See also §8.1 for the corresponding problem with unequal intervals.

Whichever method is used it is subject to the limitations and warnings at the beginning of this section and in the previous section, as the interpolation formula has to be valid before it can be used for inverse interpolation.

3.5 Backward Differences

The interpolation formula used so far is inconvenient at the end of the range of the table. For example to find e^x from Table 1 when $x = 1 \cdot 82$ we should take

$$x_0 = 1 \cdot 8,$$
$$y_0 = 6 \cdot 049647, \qquad \Delta y_0 = 0 \cdot 636247.$$

But second and higher differences are not available. To overcome this sort of difficulty a new operator ∇ is introduced, defined by the relation

$$\nabla y_0 = y_0 - y_{-1},$$

and generally

$$\nabla y_r = y_r - y_{r-1};$$

then

$$\nabla^2 y_0 = \nabla y_0 - \nabla y_{-1} = (y_0 - y_{-1}) - (y_{-1} - y_{-2})$$
$$= y_0 - 2y_{-1} + y_{-2}.$$

TABLE 3A. BACKWARD DIFFERENCES LAYOUT

y_{-3}						
∇y_{-2}						
y_{-2}	$\nabla^2 y_{-1}$					
∇y_{-1}	$\nabla^3 y_0$					
y_{-1}	$\nabla^2 y_0$	$\nabla^4 y_1$				
Δy_0	$\nabla^3 y_1$	$\nabla^5 y_2$				
y_0	$\nabla^2 y_1$	$\nabla^4 y_2$	$\nabla^6 y_3$			
∇y_1	$\nabla^3 y_2$	$\nabla^5 y_3$				
y_1	$\nabla^2 y_2$	$\nabla^4 y_3$				
∇y_2	$\nabla^3 y_3$					
y_2	$\nabla^2 y_3$					
∇y_3						
y_3						

The operator ∇ is called the *backward* difference operator whereas the operator Δ of the previous sections is called the *forward* difference operator. Corresponding to Table 3 we have Table 3A. Note that the subscripts are constant along a diagonal sloping upward from left to right.

In operator form we have

$$\nabla = 1 - E^{-1},$$

so that

$$E = (1 - \nabla)^{-1}$$

and

$$E^p = (1 - \nabla)^{-p}$$
$$= 1 + p\nabla + p(p+1)\,\nabla^2/2! + p(p+1)(p+2)\,\nabla^3/3! + \dots$$

In terms of y and its differences this algebraic identity becomes

$$y_p = y_0 + p\nabla y_0 + p(p+1)\,\nabla^2 y_0/2! +$$
$$+ p(p+1)(p+2)\,\nabla^3 y_0/3! + \dots \quad (3.5.1)$$

This is called the Gregory–Newton backward interpolation formula.

Thus to find $e^{1 \cdot 62}$ to 6 decimal places we take $x_0 = 1 \cdot 6$ and $p = 0 \cdot 2$ in Table 1. Then

$e^{1 \cdot 62}=y_{0 \cdot 2}=$	$y_0=$	$4 \cdot 953032 = 5 \cdot 053090$
$+0 \cdot 2\nabla y_0$	$+0 \cdot 2 \times 0 \cdot 471343$	
$+0 \cdot 12\nabla^2 y_0$	$+0 \cdot 12 \times 0 \cdot 044854$	
$+0 \cdot 088\nabla^3 y_0$	$+0 \cdot 088 \times 0 \cdot 004268$	
$+0 \cdot 0704\nabla^4 y_0$	$+0 \cdot 0704 \times 0 \cdot 000405$	
$+0 \cdot 0591\nabla^5 y_0$	$+0 \cdot 0591 \times 0 \cdot 000036$	

which is correct to 6 decimal places.

Exercise. Find correct to 6 decimal places sin 89° from Table 2 and $e^{1 \cdot 81}$ from Table 1 and check from Tables.

3.6 Central Differences

As explained in §3.3, the various interpolation and related formulae when applied to a general function amount to approximating the function by a polynomial which coincides with the given function at the tabular points, but usually differs from the given function at all other points. When investigating a function in the neighbourhood of x_0, the forward interpolation formula fits a polynomial at x_0, x_1, x_2, ... ; the backward interpolation formula fits a polynomial at x_0, x_{-1}, x_{-2}, ... Usually a better procedure would be to fit a polynomial at x_0, x_1 and x_{-1}, x_2 and x_{-2}, ... Of course this is not possible at the two ends of a table, but when it is possible the resulting interpolation formula is more rapidly convergent and can be described more conveniently by the use of the central difference operator δ. It is defined by the relation

$$\delta y_{\frac{1}{2}} = y_1 - y_0.$$

It is a little difficult for the beginner because of the use of the fractional subscript $\frac{1}{2}$, but a careful study of the following table, which corresponds to Tables 3 and 3A, will help a lot.

TABLE 3B. CENTRAL DIFFERENCES LAYOUT

y	δy	$\delta^2 y$	$\delta^3 y$	$\delta^4 y$	$\delta^5 y$	$\delta^6 y$
y_{-3}						
	$\delta y_{-2\frac{1}{2}}$					
y_{-2}		$\delta^2 y_{-2}$				
	$\delta y_{-1\frac{1}{2}}$		$\delta^3 y_{-1\frac{1}{2}}$			
y_{-1}		$\delta^2 y_{-1}$		$\delta^4 y_{-1}$		
	$\delta y_{-\frac{1}{2}}$		$\delta^3 y_{-\frac{1}{2}}$		$\delta^5 y_{-\frac{1}{2}}$	
y_0		$\delta^2 y_0$		$\delta^4 y_0$		$\delta^6 y_0$
	$\delta y_{\frac{1}{2}}$		$\delta^3 y_{\frac{1}{2}}$		$\delta^5 y_{\frac{1}{2}}$	
y_1		$\delta^2 y_1$		$\delta^4 y_1$		
	$\delta y_{1\frac{1}{2}}$		$\delta^3 y_{1\frac{1}{2}}$			
y_2		$\delta^2 y_2$				
	$\delta y_{2\frac{1}{2}}$					
y_3						

Note that subscripts are constant along a horizontal level and that the table does not contain, for example, δy_0 because there is no entry in the column of first differences which is at the same level as y_0. (Compare Tables 3, 3A, 3B.) The numerical entries in a difference table are the same; it is only the notation that differs. Thus if in Table 1 we take

$$1 \cdot 0 \qquad = x_0,$$

then

$$2 \cdot 718282 = y_0$$

and

$$0 \cdot 285884 = \Delta y_0 \text{ (in the forward notation)},$$
$$= \nabla y_1 \text{ (in the backward notation)},$$
$$= \delta y_{\frac{1}{2}} \text{ (in the central notation)}.$$

Observe also, to compare the three notations, that

$$\Delta^2 y_0 = 0 \cdot 030067$$
$$\nabla^2 y_0 = 0 \cdot 024617$$
$$\delta^2 y_0 = 0 \cdot 027205.$$

The relation

$$\delta y_{\frac{1}{2}} = y_1 - y_0$$

may be expressed in operator form as follows:

$$\delta y_{\frac{1}{2}} = E^{\frac{1}{2}} y_0 - E^{-\frac{1}{2}} y_0,$$

hence

$$\delta = E^{\frac{1}{2}} - E^{-\frac{1}{2}}.$$

Terms of the following type occur frequently:

$$\tfrac{1}{2}(y_0 + y_1) \quad \tfrac{1}{2}(\delta y_{-\frac{1}{2}} + \delta y_{\frac{1}{2}}) \quad \tfrac{1}{2}(\delta^2 y_0 + \delta^2 y_1) \quad \tfrac{1}{2}(\delta^3 y_{-\frac{1}{2}} + \delta^3 y_{\frac{1}{2}}) \ldots$$

Another operator, μ, called the averaging operator, is introduced to simplify the writing and handling of such terms. The formal definition is

$$\mu = \tfrac{1}{2}(E^{-\frac{1}{2}} + E^{\frac{1}{2}}),$$

and the four terms mentioned above become

$$\mu y_{\frac{1}{2}}, \qquad \mu \delta y_0, \qquad \mu \delta^2 y_{\frac{1}{2}}, \qquad \mu \delta^3 y_0$$

respectively. An expression such as $\mu \delta^2 y_0$ would mean $\tfrac{1}{2}(\delta^2 y_{-\frac{1}{2}} + \delta^2 y_{\frac{1}{2}})$, which is not available in a difference table. In the term $\mu\delta^m y_r$ if m is odd then r is integral, often 0, and if m is even then r is fractional, often $\frac{1}{2}$.

Note the following relations between the operators E, μ, and δ:

$$2\mu = E^{\frac{1}{2}} + E^{-\frac{1}{2}}, \qquad \delta = E^{\frac{1}{2}} - E^{-\frac{1}{2}},$$

$$\mu^2 = 1 + \tfrac{1}{4}\delta^2, \qquad E = 1 + \mu\delta + \tfrac{1}{2}\delta^2.$$

The first two are the definitions; the other two follow by squaring and subtracting or adding and squaring the first two relations. Note that we are able to manipulate these operators algebraically because they are commutative, i.e. $\mu\delta = \delta\mu$ (the reader should verify this statement); one has to justify carefully such manipulations.

Exercise. A function is even with respect to x_0 if $y_{-n} = y_n$ for all n, odd if $y_{-n} = -y_n$. Show that for even functions $\delta^2 y_0 = -2\delta y_{\frac{1}{2}}, \delta^4 y_0 = 2\delta^3 y_{\frac{1}{2}}, \ldots,$ and for odd functions $\delta^2 y_0 = \delta^4 y_0 = \ldots = 0$. Find from tables sinh x for $x = 0$ $(0\cdot1)$ $0\cdot4$; using these values only, construct the difference table and fill in all the entries between the horizontal line through $x_0 = 0$ and the upward diagonal through y_4. Do the same for cosh x.

3.7 Central-difference Interpolation Formulae

In this section we introduce various interpolation formulae using central differences. No proofs are given. At this stage it is more important to learn the use and structure of the formulae. In §3.10 we give various derivations of these formulae.

Newton–Stirling (S)

$$y_p = y_0 + S_1(\delta y_{\frac{1}{4}} + \delta y_{-\frac{1}{2}}) + S_2\delta^2 y_0 +$$
$$+ S_3(\delta^3 y_{\frac{1}{4}} + \delta^3 y_{-\frac{1}{2}}) + S_4\delta^4 y_0 + \ldots ,$$

where

$$S_1 = \tfrac{1}{2}p, \qquad\qquad S_4 = p^2(p^2 - 1^2)/4!,$$

$$S_2 = p^2/2!, \qquad\qquad S_5 = \tfrac{1}{2}p(p^2 - 1^2)(p^2 - 2^2)/5!,$$

$$S_3 = \tfrac{1}{2}p(p^2 - 1^2)/3!, \qquad S_6 = p^2(p^2 - 1^2)(p^2 - 2^2)/6!.$$

Note that $S_1, S_3, S_5, \ldots,$ are odd and $S_2, S_4, \ldots,$ are even with

respect to $p = 0$, e.g. $S_3(-p) = -S_3(p)$ and $S_4(-p) = S_4(p)$.
The following diagram shows the differences used in (S):

$$\delta y_{-\frac{1}{2}} \qquad\qquad \delta^3 y_{-\frac{1}{2}} \qquad\qquad \delta^5 y_{-\frac{1}{2}}$$

$$y_0 \qquad\quad \delta^2 y_0 \qquad\qquad \delta^4 y_0 \qquad\qquad \delta^6 y_0$$

$$\delta y_{\frac{1}{2}} \qquad\qquad \delta^3 y_{\frac{1}{2}} \qquad\qquad \delta^5 y_{\frac{1}{2}}$$

The differences are symmetrically distributed about an axis which
passes through the level corresponding to $p = 0$, i.e. through y_0.
The formula is suitable for interpolation for values of p near 0,
say in the range $-\frac{1}{4} < p < \frac{1}{4}$, though it does hold in the range
$-\frac{1}{2} < p < \frac{1}{2}$ and even outside this range but convergence is not
so rapid then.

Example. To find from Table 1 e^x to 6 decimal places when
$x = 1 \cdot 12$, take $x_0 = 1 \cdot 1$ so that $p = 0 \cdot 2$. Then

$$
\begin{aligned}
y_{0 \cdot 2} = y_0 &= & 3 \cdot 004166 = 3 \cdot 064854 \\
&+ S_1(\delta y_{\frac{1}{2}} + \delta y_{-\frac{1}{2}}) & + 0 \cdot 1 \times 0 \cdot 601835 \\
&+ S_2 \delta^2 y_0 & + 0 \cdot 02 \times 0 \cdot 030067 \\
&+ S_3(\delta^3 y_{\frac{1}{2}} + \delta^3 y_{-\frac{1}{2}}) & + 0 \cdot 016 \times 0 \cdot 006024.
\end{aligned}
$$

Note that as compared with forward or backward interpolation
formulae, convergence is much more rapid because the interpolation coefficients are smaller and fewer terms are computed;
4th and higher differences do not contribute to the 6th decimal
place, whereas in forward or backward formulae we need the 4th
and 5th differences to obtain the same accuracy.

Exercise. Find from Table 1 the value of $e^{1 \cdot 08}$ by taking $x_0 = 1 \cdot 1$ and
$p = -0 \cdot 2$.

Newton–Bessel (B)

$$
\begin{aligned}
y_p = y_0 &+ B_1 \delta y_{\frac{1}{2}} + B_2(\delta^2 y_0 + \delta^2 y_1) + B_3 \delta^3 y_{\frac{1}{2}} + \\
&+ B_4(\delta^4 y_0 + \delta^4 y_1) + B_5 \delta^5 y_{\frac{1}{2}} + \ldots ,
\end{aligned}
$$

where

$$B_1 = p$$
$$B_2 = \tfrac{1}{2}p(p-1)/2!$$
$$B_3 = p(p-\tfrac{1}{2})(p-1)/3!$$
$$B_4 = \tfrac{1}{2}(p+1)\,p(p-1)(p-2)/4!$$
$$B_5 = (p+1)\,p(p-\tfrac{1}{2})(p-1)(p-2)/5!.$$

Note that B_2, B_4, ..., are even and B_3, B_5, ..., are odd with respect to $p = \tfrac{1}{2}$. For example

$$B_2(0 \cdot 4) = B_2(0 \cdot 6)$$
$$B_3(0 \cdot 4) = -B_3(0 \cdot 6)$$
$$B_3(0 \cdot 5) = B_5(0 \cdot 5) = \ldots = 0.$$

The following diagram shows the differences used in (B):

y_0		$\delta^2 y_0$		$\delta^4 y_0$	
	$\delta y_{\frac{1}{2}}$		$\delta^3 y_{\frac{1}{2}}$		$\delta^5 y_{\frac{1}{2}}$
y_1		$\delta^2 y_1$		$\delta^4 y_1$	

The differences are symmetrically distributed about an axis which passes through the level corresponding to $p = \tfrac{1}{2}$.

This formula is suitable for interpolation for values of p in the range $0 < p < 1$. It is the best central-difference formula. Its excellence is due to zeros at $p = 0$, 1 for B_n $(n > 2)$ and at $p = \tfrac{1}{2}$ when n is odd and greater than 1. It can be used outside the range $0 < p < 1$ but convergence is less rapid.

Example. To find $e^{1 \cdot 14}$ from Table 1 we take $x_0 = 1 \cdot 1$ and $p = 0 \cdot 4$,

$$
\begin{aligned}
y_{0 \cdot 4} = y_0 \quad &= \quad\quad 3 \cdot 004166 = 3 \cdot 126768 \\
+ B_1 \delta y_{\frac{1}{2}} \quad & +0 \cdot 4 \times 0 \cdot 315951 \\
+ B_2(\delta^2 y_0 + \delta^2 y_1) \quad & -0 \cdot 06 \times 0 \cdot 063296 \\
+ B_3 \delta^3 y_{\frac{1}{2}} \quad & +0 \cdot 004 \times 0 \cdot 003162 \\
+ B_4(\delta^4 y_0 + \delta^4 y_1) \quad & +0 \cdot 0112 \times 0 \cdot 000632,
\end{aligned}
$$

which is correct to 6 decimal places.

A Short Table of Interpolation Coefficients

BESSEL COEFFICIENTS

p	B_2	B_3	B_4	B_5
0·1	−0·0225	0·0060	0·0039	−0·0006
0·2	−0·0400	0·0080	0·0072	−0·0009
0·3	−0·0525	0·0070	0·0097	−0·0008
0·4	−0·0600	0·0040	0·0112	−0·0004
0·5	−0·0625	0·0000	0·0117	0·0000
0·6	−0·0600	−0·0040	0·0112	0·0004
0·7	−0·0525	−0·0070	0·0097	0·0008
0·8	−0·0400	−0·0080	0·0072	0·0009
0·9	−0·0225	−0·0060	0·0039	0·0006

EVERETT COEFFICIENTS

p	E_2	F_2	E_4	F_4
0·1	−0·0285	−0·0165	0·0045	0·0033
0·2	−0·0480	−0·0320	0·0081	0·0063
0·3	−0·0595	−0·0455	0·0104	0·0089
0·4	−0·0640	−0·0560	0·0116	0·0108
0·5	−0·0625	−0·0625	0·0117	0·0117
0·6	−0·0560	−0·0640	0·0108	0·0116
0·7	−0·0455	−0·0595	0·0089	0·0104
0·8	−0·0320	−0·0480	0·0063	0·0081
0·9	−0·0165	−0·0285	0·0033	0·0045

STIRLING COEFFICIENTS

p	S_2	S_3	S_4	S_5
0·1	0·0050	−0·0082	−0·0004	0·0016
0·2	0·0200	−0·0160	−0·0016	0·0032
0·3	0·0450	−0·0227	−0·0034	0·0044
0·4	0·0800	−0·0280	−0·0056	0·0054
0·5	0·1250	−0·0312	−0·0078	0·0059
0·6	0·1800	−0·0320	−0·0096	0·0058
0·7	0·2450	−0·0297	−0·0104	0·0052
0·8	0·3200	−0·0240	−0·0096	0·0040
0·9	0·4050	−0·0142	−0·0064	0·0023

Everett's formula (E)

$$y_p = (1 - p)\, y_0 + p y_1 + E_2 \delta^2 y_0 + F_2 \delta^2 y_1 +$$
$$+ E_4 \delta^4 y_0 + F_4 \delta^4 y_1 + \ldots$$
$$E_2 = - p(p - 1)(p - 2)/3!$$
$$F_2 = (p + 1)\, p(p - 1)/3!$$
$$E_4 = - (p + 1)\, p(p - 1)(p - 2)(p - 3)/5!$$
$$F_4 = (p + 2)(p + 1)\, p(p - 1)(p - 2)/5!.$$

Note that

$$E_{2n}(p) = F_{2n}(1 - p).$$

For example,

$$E_2(0\cdot6) = F_2(0\cdot4), \qquad E_2(0\cdot4) = F_2(0\cdot6).$$

The following diagram shows the differences used in (E)

| y_0 | $\delta^2 y_0$ | $\delta^4 y_0$ |
| y_1 | $\delta^2 y_1$ | $\delta^4 y_1$ |

Formula (E) is equivalent to formula (B); it is in fact a slight rearrangement of formula (B); the same remarks about the efficiency of (B) apply to (E). The advantage of Everett's formula is that it does not use odd differences which therefore need not be tabulated.

Example. To find $e^{1\cdot16}$ from Table 1 we take $x_0 = 1\cdot1$ and $p = 0\cdot6$:

$$y_p = (1-p)\, y_0 = \qquad 0\cdot4 \times 3\cdot004166 = 3\cdot189933$$
$$+ p y_1 \qquad +0\cdot6 \times 3\cdot320117$$
$$+ E_2 \delta^2 y_0 \qquad +0\cdot056 \times 0\cdot030067$$
$$+ F_2 \delta^2 y_1 \qquad -0\cdot064 \times 0\cdot033229$$
$$+ E_4 \delta^4 y_0 \qquad +0\cdot0108 \times 0\cdot000300$$
$$+ F_4 \delta^4 y_1 \qquad +0\cdot0116 \times 0\cdot000332.$$

Bessel's or Everett's formula should be used whenever possible; Stirling's formula is suitable when $p \sim 0$. The forward and back-

3

ward interpolation formulae are used only at the two ends of a table.

All these formulae can be used for inverse interpolation in much the same way as was explained in §3.4 for the forward interpolation formula. Thus, solving the same example as in §3.4 for comparison, we require to find from Table 4 the root of

$$x^3 - 3x + 1 = 0$$

which lies between 1 and 2; using Bessel's formula we have

$$y_p = y_0 + p\delta y_{\frac{1}{2}} + B_2(\delta^2 y_0 + \delta^2 y_1) + B_3\delta^3 y_{\frac{1}{2}},$$
$$0 = -0\cdot125 + 0\cdot421p + 0\cdot186B_2 + 0\cdot006B_3.$$

The iteration for p can be put in the form

$$0\cdot421p_{n+1} = 0\cdot125 - 0\cdot186p_n(p_n - 1)/4 +$$
$$+ 0\cdot006p_n(p_n - 1)(p_n - \tfrac{1}{2})/6$$

Hence

$$p_{n+1} = 0\cdot296912 + 0\cdot11045p_n(1 - p_n) -$$
$$- 0\cdot002375p_n(1 - p_n)(\tfrac{1}{2} - p_n),$$

n	p_n
0	0
1	0·296912
2	0·319869
3	0·320848
4	0·320887.

Thus using Bessel's formula for inverse iteration requires fewer iterations than the forward formula. Using two desk machines and tabulated values of B_2, $B_3, \ldots,$ this iteration can be speeded up considerably using few or no resettings; thus we could form:

$$0\cdot421p \qquad \text{on the first machine,}$$
$$0\cdot125 - 0\cdot186B_2 - 0\cdot006B_3 \qquad \text{on the second machine.}$$

alternately adjusting p on the counter of the first machine and B_2, B_3 on the second machine until the two accumulators balance; the correct p is read off the counting register on the first machine.

In general we would form

$$y_p - B_2(\delta^2 y_0 + \delta^2 y_1) - B_3\delta^3 y_{\frac{1}{2}} \quad \text{on the first machine,}$$
$$y_0 + p\delta y_{\frac{1}{2}} + B_4(\delta^4 y_0 + \delta^4 y_1) \quad \text{on the second machine.}$$

If the registers on the desk machines are sufficiently large, this balancing act can be performed without any resetting of y and its differences; p and B_4 on the first machine and B_2, B_3 on the second machine are adjusted on the counting register.

3.8 Differentiation Formulae

One can derive a differentiation formula by differentiating with respect to p any of the interpolation formulae. From

$$x = x_0 + hp$$

we have

$$\frac{\mathrm{d}}{\mathrm{d}p} = h\frac{\mathrm{d}}{\mathrm{d}x}.$$

Thus by differentiating the forward interpolation formula we get

$$hy_p' = \Delta y_0 + \tfrac{1}{2}(2p-1)\,\Delta^2 y_0 + \tfrac{1}{6}(3p^2 - 6p + 2)\,\Delta^3 y_0 + \ldots ,$$

where the prime denotes differentiation with respect to x. Usually we are interested in the derivative at a tabular point x_0 or at a mid-interval $x_{\frac{1}{2}}$; these are obtained by putting $p = 0$ or $\tfrac{1}{2}$ and we get then

$$hy_0' = \Delta y_0 - \tfrac{1}{2}\Delta^2 y_0 + \tfrac{1}{3}\Delta^3 y_0 + \ldots , \tag{3.8.1}$$

$$hy_{\frac{1}{2}}' = \Delta y_0 + \tfrac{1}{12}\Delta^3 y_0 + \ldots$$

Similarly, by differentiating Stirling's formula once and putting $p = 0$, and by differentiating again and then putting $p = 0$, we get

$$hy_0' = \mu\delta y_0 - \tfrac{1}{6}\mu\delta^3 y_0 + \tfrac{1}{30}\mu\delta^5 y_0 - \tfrac{1}{140}\mu\delta^7 y_0 + \ldots , \tag{3.8.2}$$

$$h^2 y_0'' = \delta^2 y_0 - \tfrac{1}{12}\delta^4 y_0 + \tfrac{1}{90}\delta^6 y_0 - \tfrac{1}{560}\delta^8 y_0 + \ldots \tag{3.8.3}$$

From this last formula we derive the important formula

$$\delta^2 y_0 = h^2(y_0'' + \tfrac{1}{12}\delta^2 y_0'' - \tfrac{1}{240}\delta^4 y_0'' + \tfrac{31}{60480}\delta^6 y_0'' \ldots), \tag{3.8.4}$$

which expresses the second difference of y in terms of second and higher differences of y''. This formula may be derived by manipulation with operators and inversion of series as follows: we introduce the operator D which stands for differentiation with respect to x. Then (3.8.3) may be put in the following operator form:

$$h^2D^2 = \delta^2(1 - \tfrac{1}{12}\delta^2 + \tfrac{1}{90}\delta^4 - \ldots),$$

hence

$$\delta^2 = h^2(1 - \tfrac{1}{12}\delta^2 + \tfrac{1}{90}\delta^4 - \ldots)^{-1}D^2;$$

by expanding

$$(1 - \tfrac{1}{12}\delta^2 + \tfrac{1}{90}\delta^4 - \ldots)^{-1}$$

as a power series in δ we get

$$\delta^2 = h^2(1 + \tfrac{1}{12}\delta^2 - \tfrac{1}{240}\delta^4 + \ldots)D^2$$

which is (3.8.4) in operator form. Note that in the first step of this derivation we have assumed that the operators δ and D commute, i.e. $\delta D = D\delta$; this happens to be true for these operators, but in general, and unlike numbers, operators may not be commutative.

Finally, by differentiating Bessel's formula and then putting $p = \tfrac{1}{2}$ the coefficients of even differences become 0, so resulting in the following simple formula

$$hy'_{\frac{1}{2}} = \delta y_{\frac{1}{2}} - \tfrac{1}{24}\delta^3 y_{\frac{1}{2}} + \tfrac{3}{640}\delta^5 y_{\frac{1}{2}} - \ldots$$

The above differentiation formulae and other similar formulae can be derived alternatively by using the interesting relation which exists between E and D. Taylor's series

$$y(x_0 + h) = y_0 + hy'_0 + h^2y''_0/2! + \ldots$$

(where y_0, y'_0, $\ldots$, denote $y(x_0)$, $y'(x_0)$, $\ldots$) becomes in operator form:

$$E = 1 + hD + h^2D^2/2! + \ldots = e^{hD}.$$

Hence

$$1 + \Delta = e^{hD}$$

$$hD = \ln(1 + \Delta) = \Delta - \tfrac{1}{2}\Delta^2 + \tfrac{1}{3}\Delta^3 - \ldots,$$

which is (3.8.1) in operator form. From $E = e^{hD}$ we get

$$\delta = E^{\frac{1}{2}} - E^{-\frac{1}{2}} = e^{\frac{1}{2}hD} - e^{-\frac{1}{2}hD},$$

hence

$$\tfrac{1}{2}\delta = \sinh\left(\tfrac{1}{2}hD\right)$$

and

$$hD = 2\sinh^{-1}\left(\tfrac{1}{2}\delta\right),$$

$$h^2D^2 = 4[\sinh^{-1}\left(\tfrac{1}{2}\delta\right)]^2.$$

Let

$$f(x) = \sinh^{-1} x,$$

$$g(x) = [\sinh^{-1} x]^2.$$

An expansion of $f(x)$ in powers of x is obtained by integrating term by term the relation

$$f'(x) = (1 + x^2)^{-\frac{1}{2}} = 1 - \tfrac{1}{2}x^2 + \tfrac{3}{8}x^4 - \ldots,$$

which gives

$$f(x) = x - \tfrac{1}{6}x^3 + \tfrac{3}{40}x^5 - \ldots$$

Replacing x by $\tfrac{1}{2}\delta$ we get

$$hD = \delta - \tfrac{1}{24}\delta^3 + \tfrac{3}{640}\delta^5 - \ldots$$

To derive (3.8.2) we multiply the series for hD by the identity operator

$$\mu(1 + \tfrac{1}{4}\delta^2)^{-\frac{1}{2}} = \mu(1 - \tfrac{1}{8}\delta^2 + \tfrac{3}{128}\delta^4 - \ldots)$$

Squaring the series for hD we get

$$h^2D^2 = \delta^2 - \tfrac{1}{12}\delta^4 + \tfrac{1}{90}\delta^6 - \ldots$$

which is (3.8.3) in operator form. An alternative direct derivation of (3.8.3) is obtained by establishing the relation

$$(1 + x^2)\,g''(x) + xg'(x) - 2 = 0$$

and generally

$$(1 + x^2)\,g^{(n+2)}(x) + (2n + 1)xg^{(n+1)}(x) + n^2g^{(n)}(x) = 0$$

where $g^{(n)}(x)$ denotes the nth derivative of $g(x)$. From this relation we can quickly derive a Maclaurin series for $g(x)$, then substituting $\frac{1}{2}\delta$ for x we get (3.8.3).

While differentiation formulae are useful for deriving formulae for integration and the numerical solution of differential equations, they are not suitable for numerical evaluation of derivatives. This is because the relative error is large; the leading terms in y' and y'' are $\mu\delta/h$ and δ^2/h^2 and the relative errors in δ and δ^2 are large, and in fact the smaller is h the larger is the relative error due to cancellation in computing differences. For example, if in Table 1 we take $x_0 = 1$ then

$$y_0 = 2\cdot718282,$$

$$\delta^2 y_0 = 0\cdot027205;$$

y_0 and $\delta^2 y_0$ are both correct to 6 decimal places whereas the relative error (due to round-off) is

$$\tfrac{1}{2}\,10^{-6}/2\cdot7 \text{ in } y_0,$$

$$\tfrac{1}{2}\,10^{-6}/0\cdot027 \text{ in } \delta^2 y_0$$

and the relative error in the second derivative computed by (3.8.3) is some 100 times larger than the relative error in y_0. Graphically, we approximate the derivative by the gradient of a secant; if the interval h is large the points are widely separated; the secant is well determined but its gradient is a poor approximation to the gradient of the tangent. If the interval is small and the points are close, the secant is poorly determined but its gradient is a good approximation to the derivative.

3.9 Integration

It is convenient to make a trivial change of notation and use

$$f \quad f_0 \quad f_1 \ldots f_p$$

instead of

$$y \quad y_0 \quad y_1 \ldots y_p.$$

This is because integration formulae are often used for the

numerical solution of differential equations where y stands for the integral and f for the integrand:

$$\frac{dy}{dx} = f(x) \text{ or } f(x, y),$$

$$y = \int f(x)\, dx \text{ or } \int f(x, y)\, dx.$$

We also observe that because

$$x = x_0 + hp,$$

we have

$$\int f dx = h \int f dp$$

By integrating with respect to p the forward interpolation formula from $p = 0$ to $p = 1$ we get

$$\int_{x_0}^{x_1} f dx = h(f_0 + \tfrac{1}{2}\Delta f_0 - \tfrac{1}{12}\Delta^2 f_0 +$$
$$+ \tfrac{1}{24}\Delta^3 f_0 - \tfrac{19}{720}\Delta^4 f_0 + \dots) \quad (3.9.1)$$

Similarly by integrating with respect to p the backward interpolation formula from $p = 0$ to $p = 1$ we get

$$\int_{x_0}^{x_1} f dx = h(f_0 + \tfrac{1}{2}\nabla f_0 + \tfrac{5}{12}\nabla^2 f_0 + \tfrac{3}{8}\nabla^3 f_0 +$$
$$+ \tfrac{251}{720}\nabla^4 f_0 + \dots). \quad (3.9.2)$$

By integrating the backward interpolation formula from $p = -1$ to $p = 0$ and then advancing the subscript by 1, or by multiplying the right-hand side of (3.9.2) by the identity operator $(1 - \nabla)$ E we get

$$\int_{x_0}^{x_1} f dx = h(f_1 - \tfrac{1}{2}\nabla f_1 - \tfrac{1}{12}\nabla^2 f_1 - \tfrac{1}{24}\nabla^3 f_1 -$$
$$- \tfrac{19}{720}\nabla^4 f_1 - \dots). \quad (3.9.3)$$

Formulae (3.9.2) and (3.9.3) are used for the numerical solution of differential equations. Formula (3.9.2) is an extrapolation formula: it uses the ordinates at x_0, x_{-1}, x_{-2}, ..., to find the integral up to x_1; for this reason it is called a *predictor* whereas (3.9.3) is called a *corrector*; as it makes use of f_1 one expects it to be more accurate and in fact its coefficients are smaller and therefore it is more rapidly convergent than the predictor.

Integrating Bessel's formula from $p = 0$ to $p = 1$ we get

$$\int_{x_0}^{x_1} f \mathrm{d}x = h(\mu f_{\frac{1}{2}} - \tfrac{1}{12}\mu\delta^2 f_{\frac{1}{2}} + \tfrac{11}{720}\mu\delta^4 f_{\frac{1}{2}} -$$
$$- \tfrac{191}{60480}\mu\delta^6 f_{\frac{1}{2}} + \dots). \quad (3.9.4)$$

Like other central-difference formulae, this formula is much more powerful than the integration formulae using forward or backward differences but it cannot be used at the two ends of a table.

Next we derive two useful formulae for integration over several intervals. Let

$$I = \int_{x_0}^{x_n} f \mathrm{d}x.$$

The simplest approximation to I is given by the trapezoidal rule: $I \sim hT$ where

$$T = \tfrac{1}{2}f_0 + f_1 + f_2 + \dots + f_{n-1} + \tfrac{1}{2}f_n.$$

If in (3.9.4) we replace

$$\mu f_{\frac{1}{2}} \text{ by } \tfrac{1}{2}(f_0 + f_1)$$

and

$$\delta f_{\frac{1}{2}} \text{ by } f_1 - f_0$$

the expression inside the bracket becomes

$$\tfrac{1}{2}(f_0 + f_1) + (-\tfrac{1}{12}\mu\delta + \tfrac{11}{720}\mu\delta^3 - \dots)(f_1 - f_0).$$

Applying (3.9.4) in this form to the intervals

$$(x_0 \ x_1), \ (x_1 \ x_2), \ (x_2 \ x_3), \ \dots , \ (x_{n-1} \ x_n)$$

and adding we get
$$I = hT + hC,$$

where T is as above and

$$C = -\tfrac{1}{12}\mu\delta(f_n - f_0) + \tfrac{11}{720}\mu\delta^3(f_n - f_0) - \ldots \qquad (3.9.5)$$

When x_0, x_n are the first and last entries in a table, a more useful expression for C is given by the Gregory formula:

$$C = \tfrac{1}{12}(\Delta f_0 - \nabla f_n) - \tfrac{1}{24}(\Delta^2 f_0 + \nabla^2 f_n) +$$

$$+ \tfrac{19}{720}(\Delta^3 f_0 - \nabla^3 f_n) - \tfrac{3}{160}(\Delta^4 f_0 + \nabla^4 f_n) \ldots \qquad (3.9.6)$$

To derive this formula we introduce the operator $\dfrac{1}{D}$ which stands for indefinite integration; $\dfrac{1}{D} f$ is an indefinite integral and contains an arbitrary constant of integration. An expression such as $\dfrac{\Delta}{D} f_0$ is ambiguous as the operators Δ and $\dfrac{1}{D}$ are not commutative; we use the expression $\dfrac{\Delta}{D} f_0$ to mean $\dfrac{1}{D} f_1 - \dfrac{1}{D} f_0$ with the same constant of integration in the two integrals so that

$$\frac{\Delta}{D} f_0 = \frac{1}{D} f_1 - \frac{1}{D} f_0 = \int_{x_0}^{x_1} f \, \mathrm{d}x = h \int_0^1 f \, \mathrm{d}p.$$

The other interpretation

$$\frac{\Delta}{D} f_0 = \frac{1}{D} \Delta f_0 = \frac{1}{D} (f_1 - f_0)$$

is not defined.

From the relation
$$hD = \ln (1 + \Delta)$$

established in the previous chapter we get

$$\frac{\Delta}{h\mathrm{D}} = \frac{\Delta}{\ln(1+\Delta)} = \frac{\Delta}{\Delta - \frac{1}{2}\Delta^2 + \frac{1}{3}\Delta^3 - \ldots}$$
$$= 1 + \frac{1}{2}\Delta - \frac{1}{12}\Delta^2 + \frac{1}{24}\Delta^3 - \ldots$$

which is (3.9.1) in operator form.

This last relation can be put in the form

$$\frac{1}{h\mathrm{D}} = \Delta^{-1} + \frac{1}{2} - \frac{1}{12}\Delta + \frac{1}{24}\Delta^2 - \ldots$$

The operator Δ^{-1} needs careful definition; like $\frac{1}{\mathrm{D}}$ it contains an arbitrary constant (of summation). From

$$\Delta f_r = f_{r+1} - f_r$$

we get

$$f_r = \Delta^{-1} f_{r+1} - \Delta^{-1} f_r.$$

By adding this identity for $r = 0, 1, 2, \ldots, n-1$, we get

$$\Delta^{-1} f_n - \Delta^{-1} f_0 = f_0 + f_1 + f_2 + \ldots + f_{n-1}.$$

Similar results can be established for the backward difference operator ∇;

$$\mathrm{E} = (1 - \nabla)^{-1},$$
$$h\mathrm{D} = -\ln(1 - \nabla),$$
$$\frac{1}{h\mathrm{D}} = \nabla^{-1} - \frac{1}{2} - \frac{1}{12}\nabla - \frac{1}{24}\nabla^2 - \ldots,$$

which is (3.9.3) in operator form. Moreover

$$\nabla^{-1} f_n - \nabla^{-1} f_0 = f_1 + f_2 + \ldots + f_n.$$

The relation between Δ^{-1} and ∇^{-1} is derived from the relation

$$\Delta = \mathrm{E}\nabla;$$

hence

$$\nabla^{-1} = \mathrm{E}\Delta^{-1} \text{ and } \nabla^{-1} f_n = \Delta^{-1} f_{n+1}.$$

Finally we express the definite integral in the form

$$\int_{x_0}^{x_n} f \mathrm{d}x = \frac{1}{h\mathrm{D}} f_n - \frac{1}{h\mathrm{D}} f_0,$$

where

$$\frac{1}{h\mathrm{D}} f_n = \nabla^{-1} f_n - (\tfrac{1}{2} + \tfrac{1}{12}\nabla + \tfrac{1}{24}\nabla^2 + \dots) f_n$$

and

$$\frac{1}{h\mathrm{D}} f_0 = \Delta^{-1} f_0 + (\tfrac{1}{2} - \tfrac{1}{12}\Delta + \tfrac{1}{24}\Delta^2 - \dots) f_0.$$

Substituting these values in $\frac{1}{h\mathrm{D}} f_n - \frac{1}{h\mathrm{D}} f_0$ the leading terms become

$$\nabla^{-1} f_n - \Delta^{-1} f_0 - \tfrac{1}{2}(f_n + f_0)$$
$$= \Delta^{-1} f_{n+1} - \Delta^{-1} f_0 - \tfrac{1}{2}(f_n + f_0)$$
$$= (f_0 + f_1 + f_2 + \dots + f_n) - \tfrac{1}{2}(f_0 + f_n)$$
$$= \tfrac{1}{2} f_0 + f_1 + \dots + f_{n-1} + \tfrac{1}{2} f_n$$
$$= T.$$

The remaining terms in $\frac{1}{h\mathrm{D}} f_n - \frac{1}{h\mathrm{D}} f_0$ simplify to the expression C as given by (3.9.6).

We give four useful integration formulae and derive the first two of them.

$$\int_{x_0}^{x_2} f \mathrm{d}x = \tfrac{1}{3}h(f_0 + 4f_1 + f_2) + R, \qquad\qquad (3.9.7)$$

"Simpson's rule".

$$R = \tfrac{-1}{90} h^5 f^{\mathrm{iv}}(u) \sim \tfrac{-1}{90} h \Delta^4 f_0, \text{ where } x_0 \leqslant u \leqslant x_2.$$

$$\int_{x_0}^{x_4} f \mathrm{d}x = \tfrac{4}{3}h(2f_1 - f_2 + 2f_3) + R, \qquad\qquad (3.9.8)$$

$$R = \tfrac{28}{90} h^5 f^{\mathrm{iv}}(u) \sim \tfrac{28}{90} h \Delta^4 f_0, \text{ where } x_0 \leqslant u \leqslant x_4.$$

$$\int_{x_0}^{x_3} f\mathrm{d}x = \tfrac{3}{8}h(f_0+3f_1+3f_2+f_3)+R, \qquad\qquad (3.9.9)$$

"Three-eighths rule".

$$R = \tfrac{-3}{80}h^5 f^{\mathrm{iv}}(u), \qquad\qquad \text{where } x_0 \leqslant u \leqslant x_3.$$

$$\int_{x_0}^{x_6} f\mathrm{d}x = \tfrac{3}{10}h(f_0+5f_1+f_2+6f_3+f_4+5f_5+f_6)+R, \qquad (3.9.10)$$

"Weddle's formula".

$$R = -\tfrac{1}{140}h^7 f^{\mathrm{vi}}(u), \qquad\qquad \text{where } x_0 \leqslant u \leqslant x_6.$$

In the above equations f^{iv} denotes the 4th derivative and f^{vi} the 6th derivative.

Simpson's rule is usually applied over several intervals in the form:

$$\int_{x_0}^{x_{2n}} f\mathrm{d}x \sim \tfrac{1}{3}h(f_0+4f_1+2f_2+4f_3+2f_4+ \ldots +4f_{2n-1}+f_{2n}).$$

To derive (3.9.7) we integrate the forward interpolation formula

$$f_p = f_0 + p\Delta f_0 + \tfrac{1}{2}(p^2 - p)\,\Delta^2 f_0 + \tfrac{1}{6}(p^3 - 3p^2 + 2p)\,\Delta^3 f_0 + \\ + \tfrac{1}{24}(p^4 - 6p^3 + 11p^2 - 6p)\,\Delta^4 f_0 + \ldots$$

using the relation

$$\int_{x_0}^{x_2} f\mathrm{d}x = h\int_{0}^{2} f_p\mathrm{d}p.$$

The integral of the first three terms gives

$$h(2f_0 + 2\Delta f_0 + \tfrac{1}{3}\Delta^2 f_0) = \tfrac{1}{3}h(f_0 + 4f_1 + f_2).$$

The integral of the term in $\Delta^3 f_0$ is 0. The integral of the remaining terms is the remainder R. We consider now in general the remainder term in interpolation and integration formulae; for, as mentioned in §3.3, a study of the remainder term is necessary to establish the validity of various approximations in interpolation, integration, or differentiation.

From the basic mean value theorem of the Differential Calculus we have

$$\Delta f_0 = f_1 - f_0 = hf'(u), \quad \text{where } x_0 \leqslant u \leqslant x_1.$$

By repeated application of this result we obtain successively

$$\Delta^2 f_0 = h^2 f''(u), \quad \text{where } x_0 \leqslant u \leqslant x_2,$$

. . . .

$$\Delta^4 f_0 = h^4 f^{\text{iv}}(u), \quad \text{where } x_0 \leqslant u \leqslant x_4.$$

Thus if the forward interpolation formula is truncated after 4 terms the leading term in the remainder is

$$\binom{p}{4} h^4 f^{\text{iv}}(u). \tag{a}$$

Similarly, the leading term in R, the remainder in the integration formula, is

$$h \int_0^2 \binom{p}{4} \Delta^4 f_0 \mathrm{d}p = \frac{-h}{90} \Delta^4 f_0 = -\frac{h^5}{90} f^{\text{iv}}(u). \tag{b}$$

In fact one can show that when truncating the interpolation formula after 4 terms one can choose u so that the expression (a) is equal to the whole remainder, not just the leading term; similarly one can choose u so that (b) is equal to the whole remainder in the integration formula (3.9.7). The proofs are beyond the scope of this book. The form (a) of the remainder for interpolation formulae is a special case of the result given in §8.1 for unequal intervals.

Expressions for the remainder such as (a) or (b) are valid only if the 4th derivative of $f(x)$ exists throughout the range $(x_0 \ x_4)$, which implies the continuity of $f(x)$ and the existence and continuity of its first three derivatives; the interpolation and integration formulae and the expressions for their remainder are not valid if $f(x)$ has a singularity in the range investigated, and not very useful if $f(x)$ is nearly singular as the remainder term can be large and the approximation will be poor. The results explained above are a special instance of a generally useful guide: when a formula is obtained by truncating a rapidly convergent infinite series, a good estimate of the error is the first non-zero term neglected. Thus it was suggested in §3.3 that using the approximation

$$f_p = f_0 + p\Delta f_0 + \binom{p}{2} \Delta^2 f_0 + \binom{p}{3} \Delta^3 f_0$$

is equivalent to approximating the function $f(x)$ by a cubic polynomial, but this does not give any indication of the magnitude of the truncation error R; in this case

$$R = \binom{p}{4} h^4 f^{\text{iv}}(u) \sim \binom{p}{4} \Delta^4 f_0.$$

We now return to consider some aspects of the practical use of Simpson's rule. The efficiency of this rule is due to the fact that although it seems to go only up to the term in Δ^2, in fact the term in Δ^3 does not contribute to the integral. When using an integration formula one can improve accuracy by taking the interval h sufficiently small, but this requires several evaluations of the integrand; the object of an integration program is to attain the desired accuracy with as few evaluations of the integrand as possible. Simpson's rule is particularly suitable for this process of choosing the appropriate size of interval for integration. One evaluates

$$I = \int_a^b f(x)\,\mathrm{d}x$$

from the formula

$$I = \tfrac{1}{3}h(f_0 + f_{2n} + 2A + 4B),$$

where

h = interval of integration,

$2n$ = number of intervals, so that $b - a = 2nh$,

$A = f_2 + f_4 + \ \ldots \ + f_{2n-2}$,

$B = f_1 + f_3 + \ \ldots \ + f_{2n-1}$.

If now we repeat the procedure with n replaced by $m = 2n$ and h by $h' = \frac{1}{2}h$ we get

$$I' = \tfrac{1}{3}h'(f_0 + f_{2m} + 2A' + 4B'),$$

where

$$f_{2m} = f_{2n},$$
$$A' = A + B,$$
$$B' = f_1' + f_3' + \ldots + f_{2m-1}'.$$

Hence

$$I' = I + E,$$

where the correction E is given by the formula

$$E = \tfrac{1}{3}h(4B' - 2B) - \tfrac{1}{2}I$$

The program will then be as follows:

(1) Set $n = 1$, $h = \frac{1}{2}(b - a)$. Evaluate f_0, f_1, f_2 and set $I = \frac{1}{3}h(f_0 + 4f_1 + f_2)$ and $B' = f_1$.

(2) Replace n by $2n$, h by $\frac{1}{2}h$, B by B', and evaluate the mid-ordinates f_1, f_3, $\ldots$, f_{2n-1} for the new interval h. Set $B' = f_1 + f_3 + \ldots + f_{2n-1}$.

(3) Calculate the correction $E = \frac{1}{3}h(4B' - 2B) - \frac{1}{2}I$ and replace I by $I + E$.

(4) The integration is terminated if E is numerically less than a specified accuracy parameter, otherwise repeat from step 2.

If $f(x)$ is very rapidly varying, e.g. if it is very steep or oscillatory, it will be found that the interval will be halved several times before the integration terminates. If $f(x)$ is slowly varying and does not change very much in the range (ab) then maybe a few intervals only will be enough. It follows that the procedure described above should not be used if, for example $f(x)$ is rapidly varying in one part of the range (ab) and slowly varying in the other part. Thus to find the area under the curve in Fig. 2 from P to R it would not be economical to take the interval (ab) from P to R. First we might take (ab) to be from P to Q and then we take (ab) to be from Q to R.

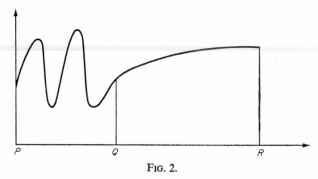

Fig. 2.

Next we derive (3.9.8). This can be done by integrating Stirling's formula, which can be put in the form

$$f_p = f_0 + S_1(\delta f_{\frac{1}{2}} + \delta f_{-\frac{1}{2}}) + S_2\delta^2 f_0 +$$
$$+ S_3(\delta^3 f_{\frac{1}{2}} + \delta^3 f_{-\frac{1}{2}}) + S_4\delta^4 f_0 + \dots ,$$

where

$$S_1 = \tfrac{1}{2}p, \qquad S_3 = \tfrac{1}{12}p(p^2 - 1),$$
$$S_2 = \tfrac{1}{2}p^2, \qquad S_4 = \tfrac{1}{24}p^2(p^2 - 1).$$

Let

$$I = \int_{x_{-2}}^{x_2} f(x)\,\mathrm{d}x = h \int_{-2}^{2} f_p\,\mathrm{d}p.$$

We integrate term by term. The integrals of S_1, S_3 and all the odd coefficients are zero. The rest gives

$$\frac{1}{h}I = 4f_0 + \tfrac{8}{3}\delta^2 f_0 + \tfrac{28}{90}\delta^4 f_0 + \dots$$
$$= \tfrac{4}{3}(2f_{-1} - f_0 + 2f_2) + \tfrac{28}{90}\delta^4 f_0 + \dots$$

which is (3.9.8) applied at x_0 instead of x_2. Simpson's rule could also have been derived by integrating Stirling's formula between x_{-1} and x_1. As in the case of Simpson's rule we do not give a rigorous derivation of the remainder term. We have only deduced the leading term in this error. We observe that it is 28 times larger than the error in Simpson's rule. Formula (3.9.8) is also unsatis-

factory because one of the coefficients is negative which will involve some cancellation though not severe, usually without corresponding cancellation in any round-off or approximation errors
in the integrand. This formula, like (3.9.3), is of the open or
predictor type.

In general integration formulae are numerically stable. Integration is a smoothing process. At the end of the section on differentiation it was observed that differentiation formulae are numerically unstable because of loss of relative accuracy due to cancellation in the leading term. Conversely integration is good because
summation reduces the relative effect of round-off or other
errors. When integrating over several intervals the leading term is
of the form

$$a_0 f_0 + a_1 f_1 + \ldots + a_n f_n.$$

Usually all the coefficients and all the terms in this expression are
of the same sign and there is no cancellation, whereas rounding
errors are not usually of the same sign and they cancel out because
of their randomness; the statistical sum of rounding errors represents a smaller relative error in the accumulated sum than the
relative round-off error in each term.

3.10 Derivation of the Central-difference Interpolation Formulae

We recall the identities of §3.6:

$$\delta = E^{\frac{1}{2}} - E^{-\frac{1}{2}},$$
$$\mu = \tfrac{1}{2}(E^{\frac{1}{2}} + E^{-\frac{1}{2}}),$$
$$\mu^2 = 1 + \tfrac{1}{4}\delta^2,$$
$$E = 1 + \mu\delta + \tfrac{1}{2}\delta^2.$$

From the first two we have

$$E = 1 + \delta E^{\frac{1}{2}}$$
$$E^{\frac{1}{2}} = \mu + \tfrac{1}{2}\delta$$

To derive Bessel's formula we express the operator E^p in the form:

$$E^p = (1 + \delta E^{\frac{1}{2}})^p$$

$$= 1 + \left\{ p\delta + \binom{p}{2} \delta^2 E^{\frac{1}{2}} + \binom{p}{3} \delta^3 E + \binom{p}{4} \delta^4 E^{1\frac{1}{2}} + \ldots \right\} E^{\frac{1}{2}}.$$

In the expression inside the big brackets we replace

$$E^{\frac{1}{2}} \quad \text{by } (\mu + \tfrac{1}{2}\delta),$$
$$E \quad \text{by } (\mu + \tfrac{1}{2}\delta)^2,$$
$$E^{1\frac{1}{2}} \text{ by } (\mu + \tfrac{1}{2}\delta)^3,$$
$$\ldots\ldots\ldots\ldots\ldots$$

then

$$\mu^2 \text{ by } 1 + \tfrac{1}{4}\delta^2,$$
$$\mu^3 \text{ by } \mu(1 + \tfrac{1}{4}\delta^2),$$
$$\mu^4 \text{ by } (1 + \tfrac{1}{4}\delta^2)^2,$$
$$\mu^5 \text{ by } \mu(1 + \tfrac{1}{4}\delta^2)^2$$
$$\ldots\ldots\ldots\ldots\ldots$$

After collecting similar terms we get

$$E^p = 1 + \left\{ p\delta + \binom{p}{2} \mu\delta^2 + \frac{p(p-1)(p-\frac{1}{2})}{3!} \delta^3 + \ldots \right\} E^{\frac{1}{2}},$$

which is Bessel's formula in operator form.

Everett's formula is derived from Bessel's by expressing odd differences in terms of even differences of lower order, e.g.

$$\delta y_{\frac{1}{2}} = y_1 - y_0,$$
$$\delta^3 y_{\frac{1}{2}} = \delta^2 y_1 - \delta^2 y_0,$$
$$\ldots\ldots\ldots\ldots\ldots\ldots$$

and simplifying we get

$$E_{2r} = B_{2r} - B_{2r+1},$$
$$F_{2r} = B_{2r} + B_{2r+1}.$$

Stirling's formula can be derived in a manner similar to Bessel's by expanding the right-hand side of the identity

$$E^p = (1 + \mu\delta + \tfrac{1}{2}\delta^2)^p.$$

A convenient method of deriving any of the central-difference interpolation formulae involves expressing everything in terms of Δ and equating coefficients; this method has the advantage that it gives the general term. For example, Stirling's formula can be put in the form

$$E^p = (1 + \Delta)^p = 1 + S_1(\delta_{\frac{1}{2}} + \delta_{-\frac{1}{2}}) + S_2\delta_0^2 +$$
$$+ S_3(\delta_{\frac{1}{2}}^3 + \delta_{-\frac{1}{2}}^3) + S_4\delta_0^4 + \ldots$$

Now

$$\delta y_{\frac{1}{2}} = \Delta y_0,$$
$$\delta y_{-\frac{1}{2}} = \Delta y_{-1} = \Delta E^{-1} y_0 = \frac{\Delta}{1 + \Delta} y_0,$$

hence

$$\delta_{\frac{1}{2}} + \delta_{-\frac{1}{2}} = \Delta + \frac{\Delta}{1 + \Delta}.$$

Similarly

$$\delta^2 y_0 = \Delta^2 y_{-1} = \frac{\Delta^2}{1 + \Delta} y_0,$$

$$\delta^3 y_{\frac{1}{2}} + \delta^3 y_{-\frac{1}{2}} = \Delta^3 y_{-1} + \Delta^3 y_{-2} = \frac{\Delta^3}{1 + \Delta} y_0 + \frac{\Delta^3}{(1 + \Delta)^2} y_0,$$

$$\delta^4 y_0 = \Delta^4 y_{-2} = \frac{\Delta^4}{(1 + \Delta)^2} y_0.$$

Stirling's formula becomes

$$(1 + \Delta)^p = 1 + \left(\Delta + \frac{\Delta}{1 + \Delta}\right) S_1 + \frac{\Delta^2}{1 + \Delta} S_2 +$$

$$+ \left(\frac{\Delta^3}{1 + \Delta} + \frac{\Delta^3}{(1 + \Delta)^2}\right) S_3 + \frac{\Delta^4}{(1 + \Delta)^2} S_4 + \ldots$$

Multiplying by $1 + \Delta$ we get

$$(1 + \Delta^p)^{+1} = (1 + \Delta) + (2\Delta + \Delta^2) S_1 + \Delta^2 S_2 +$$
$$+ \left(\Delta^3 + \frac{\Delta^3}{1 + \Delta}\right) S_3 + \frac{\Delta^4}{1 + \Delta} S_4 + \ldots$$

Equating coefficients of Δ we get

$$\binom{p + 1}{1} = 1 + 2S_1 \text{ hence } S_1 = \tfrac{1}{2} \binom{p + 1}{1} - \tfrac{1}{2} = \tfrac{1}{2}p.$$

Equating coefficients of Δ^2 we get

$$S_2 + S_1 = \binom{p + 1}{2}, \text{ hence } S_2 = \binom{p + 1}{2} - \tfrac{1}{2}p = \tfrac{1}{2}p^2.$$

Multiplying by $1 + \Delta$ again we get

$$(1 + \Delta)^{p+2} = (1 + \Delta)^2 + (2\Delta + \Delta^2)(1 + \Delta) S_1 +$$
$$+ (1 + \Delta) \Delta^2 S_2 + (2\Delta^3 + \Delta^4) S_3 + \Delta^4 S_4 + \ldots$$

Equating coefficients of Δ^3 we get

$$S_1 + S_2 + 2 S_3 = \binom{p + 2}{3}, \text{ hence}$$

$$S_3 = \tfrac{1}{2} \binom{p + 1}{3},$$

equating coefficients of Δ^4 we get

$$S_4 = \binom{p + 2}{4} - \tfrac{1}{2} \binom{p + 1}{3},$$

and generally we see that

$$S_{2n+1} = \tfrac{1}{2} \binom{p + n}{2n + 1},$$

$$S_{2n+2} = \binom{p + n + 1}{2n + 2} - \tfrac{1}{2} \binom{p + n}{2n + 1}.$$

Exercise. Use this method to derive *Gauss's forward formula:*

$$y_p = y_0 + \binom{p}{1} \delta y_{\frac{1}{2}} + \binom{p}{2} \delta^2 y_0 + \binom{p+1}{3} \delta^3 y_{\frac{1}{2}} +$$
$$+ \binom{p+1}{4} \delta^4 y_0 + \ldots , \qquad (G_1)$$

the general terms being

$$\binom{p+n-1}{2n-1} \delta^{2n-1} y_{\frac{1}{2}} \text{ and } \binom{p+n-1}{2n} \delta^{2n} y_0.$$

By replacing

$$\delta^{2n-1} y_{\frac{1}{2}} \text{ by } \delta^{2n-1} y_{-\frac{1}{2}} + \delta^{2n} y_0,$$

and using the identity

$$\binom{p}{r} + \binom{p}{r-1} = \binom{p+1}{r}$$

derive from (G₁) *Gauss's backward formula:*

$$y_p = y_0 + \binom{p}{1} \delta y_{-\frac{1}{2}} + \binom{p+1}{2} \delta^2 y_0 + \binom{p+1}{3} \delta^3 y_{-\frac{1}{2}} +$$
$$+ \binom{p+2}{4} \delta^4 y_0 + \ldots \qquad (G_2)$$

Derive Stirling's formula from the mean of (G₁) and (G₂). Derive Bessel's formula from (G₁) by replacing

$$\delta^{2n} y_0 \text{ by } \tfrac{1}{2}(\delta^{2n} y_0 + \delta^{2n} y_1) - \tfrac{1}{2} \delta^{2n+1} y_{\frac{1}{2}},$$

and hence establish the general formula for the Bessel coefficients:

$$B_{2n} = \tfrac{1}{2} \binom{p+n-1}{2n} \text{ and } B_{2n+1} = \binom{p+n}{2n+1} - \tfrac{1}{2} \binom{p+n-1}{2n}$$

3.11 Concluding Remarks

Differences are not always suitable for digital computers because they may need too much storage space. Interpolation and other formulae in terms of finite differences are avoided by expressing them in terms of ordinates as in formulae (3.9.7)–(3.9.10). Thus (3.9.4) becomes

$$\int_0^1 f(x) \, dx = \frac{h}{1440} (11f_{-2} - 93f_{-1} + 802f_0 +$$
$$+ 802f_1 - 93f_2 + 11f_3).$$

One limitation of finite difference methods is that they apply only when the function is given at equal intervals; in Chapter 8 we give various methods for approximations, interpolation, and integration for unequal intervals. Another limitation of interpolation formulae, whether they are in terms of equal intervals or not, is that they are of *local* nature. Interpolation formulae should not be used outside the principal interval if it can be avoided. For example, Newton's interpolation formula

$$y_p = y_0 + \binom{p}{1}_\Delta y_0 + \binom{p}{2}^2_{\Delta^2} y_0 + \ldots + \binom{p}{n}^n_{\Delta^n} y_0$$

truncated after $n + 1$ terms is most efficient when $p \sim \frac{1}{2}n$ and gets less and less accurate when p is further away from $\frac{1}{2}n$. Similarly, Stirling's formula should be used for $-\frac{1}{2} < p < \frac{1}{2}$, and Bessel's for $0 < p < 1$. To get better results from an interpolation formula one should make the intervals smaller rather than take more terms of the formula into account. Thus there is a need for a method of approximating a given complicated function by another function which is easily computable so that the approximation is equally good over a wide range, possibly the whole range of the function. This is achieved in Chapter 8 by means of least square methods, curve fitting, and orthogonal polynomials; these methods are of *global* rather than local nature.

EXERCISES

1. The following table of exact values contains an error. Correct it, then use Newton's formula to interpolate the exact value when $x = 0 \cdot 42$.

x	0	0·1	0·2	0·3	0·4	0·5	0·6	0·7	0·8	0·9
y	10·30	10·70	11·04	11·26	11·30	11·01	10·60	9·74	8·46	6·70

(B.C.)

[*Hint*: the function is $y = 10 \cdot 3 + 4 \cdot 1x - 10x^3$.]

2. Given the following table for x and y, find x correct to 2 decimal places when $y = 0$:

x	1	2	3	4	5
y	-8	-1	18	55	116

What is the least degree of a polynomial in x which has the values given in the table? What will be the value of this polynomial at $x = 6$?

(B.C.)

[*Hint:* the function is $x^3 - 9$.]

3. Find, from the table below, the value of y when $x = 0\cdot5$:

x	0	0·2	0·4	0·6	0·8	1
y	0·6989700	0·7075702	0·7160033	0·7242759	0·7323938	0·7403627

(B.C.)

[*Ans.* 0·7201593.]

4. Locate the error in the following table and replace it by an improved value; then find x to 4 significant figures when $y = 0$:

x	0	0·1	0·2	0·3	0·4
y	$-1\cdot20280$	$-1\cdot07500$	$-0\cdot91228$	$-0\cdot69202$	$-0\cdot38296$

x	0·5	0·6	0·7	0·8	0·9	1
y	0·05000	0·65000	1·46882	2·56808	4·02044	5·91080

(B.C.)

5. Show that a curve $y = f(x)$, where $f(x)$ is a 4th degree polynomial, can be drawn through the points given by the following table. Find y exactly when $x = 1\cdot2$.

x	-1	0	1	2	3	4	5
y	23	13	3	1	34	148	408

(U.L.)

6. The values of a low degree polynomial are given in the table below. It is suspected that there is a transposition error in one of the values. By differencing locate and correct the error and find $f(2\cdot5)$.

x	2	3	4	5	6	7	8	9	10
$f(x)$	15	40	85	165	259	400	585	820	1111

(U.L.)

7. Find, from the following table, y and y' when $x = 1\cdot45$:

x	1	1·2	1·4	1·6	1·8	2
y	0	$-0\cdot112$	$-0\cdot016$	0·336	0·992	2

(U.L.)

8. The following table gives the coordinates of points on a certain polynomial curve. Calculate the radius of curvature at the point $x = 0\cdot6$.

x	0	0·2	0·4	0·6	0·8	1·0	1·2
y	0·710	1·175	1·811	2·666	3·801	5·292	7·232

(U.L.)

9. Find from the table below, x correct to 2 decimal places, for which y is a maximum and determine this value of y.

x	1·5	1·6	1·7	1·8	1·9
y	0·9750	0·9957	0·9166	0·7385	0·4630

(B.C.)

[*Ans.* 1·57, 1·0000.]

10. Tabulate $\dfrac{1}{1 + x^2}$ for $x = 0\,(0\cdot1)\,1\cdot0$. By the use of this table and suitable numerical integration formulae, tabulate

$$\arctan x = \int_0^x \frac{1}{1 + t^2}\,dt$$

for $x = 0\,(0\cdot1)\,1\cdot0$ and check from tables.

CHAPTER 4

Recurrence Relations and Algebraic Equations

4.1 Recurrence Relations: Difference Equations

A sequence of quantities

$$y_0, y_1, y_2, y_3, \cdots$$

are said to satisfy a *linear recurrence relation of order n* if each y can be expressed and computed as a linear combination of the n immediately preceding y's. For example

$$y_{m+2} = ay_{m+1} + by_m \qquad (b \neq 0) \qquad (4.1.1)$$

is a linear recurrence relation of order 2. Given the initial quantities y_0 and y_1 we can compute

$$y_2 = ay_1 + by_0,$$
$$y_3 = ay_2 + by_1,$$
$$\cdots \text{ etc.}$$

In general, the coefficients a, b vary and depend upon m, but we confine ourselves to the simpler case with *constant coefficients*.

Recurrence relations are often expressed as *difference equations*. Consider y_m to be a function of m given at intervals $h = 1$. Then

$$y_{m+1} = \mathrm{E}y_m = (1 + \Delta)\, y_m,$$
$$y_{m+2} = \mathrm{E}^2 y_m = (1 + \Delta)^2\, y_m,$$
$$\cdots \text{ etc.}$$

The recurrence relation can then be re-expressed in terms of y_m

and its differences. Thus (4.1.1) becomes

$$(1 + 2\Delta + \Delta^2)\, y_m = a(1 + \Delta)\, y_m + by_m$$

or

$$\Delta^2 y_m + (2 - a)\, \Delta y_m + (1 - a - b)\, y_m = 0,$$

and in general a recurrence relation of order n can be expressed in the form of a difference equation involving

$$y_m,\ \Delta y_m,\ \Delta^2 y_m,\ \ldots,\ \Delta^n y_m.$$

The general linear recurrence relation of order n with constant coefficients can be put in the form

$$y_{m+n} + a_1 y_{m+n-1} + a_2 y_{m+n-2} +$$
$$+ \ldots + a_n y_m = 0 \qquad (a_n \neq 0). \quad (4.1.2)$$

To "solve" this recurrence relation means to find an explicit formula for y_m which does not involve the preceding y's. Several methods for solving linear recurrence relations are almost identical with those for solving linear differential equations.

Try a solution of the form

$$y_r = Ax^r \qquad A \neq 0$$

Substituting in (4.1.2) and omitting a common factor Ax^n, we get

$$x^n + a_1 x^{n-1} + \ldots + a_n = 0.$$

If the roots of this equation are distinct, say $(x_1, x_2, \ldots, x_n)$, then each root gives a different solution and since the equation is linear, the different solutions are additive and the general solution is of the form

$$y_m = A_1 x_1^m + A_2 x_2^m + \ldots + A_n x_n^m.$$

Thus the general solution has n arbitrary constants which are determined by the initial values $(y_0, y_1, \ldots, y_{n-1})$. This form of the general solution is valid even if the roots x_i are complex, but they must be distinct.

It is worthwhile to express the recurrence relation and the

method of solution in terms of operators. Let $f(x)$ be defined as the polynomial

$$f(x) = x^n + a_1 x^{n-1} + a_2 x^{n-2} + \ldots + a_n.$$

If we replace y_{m+1}, y_{m+2}, $\ldots$ by Ey_m, $E^2 y_m, \ldots$, then (4.1.2) becomes

$$f(E) y_m = 0.$$

Thus $f(E)$ is an operator which annihilates y_m, and $(x_1, x_2, \ldots, x_n)$ are the roots of the equation

$$f(E) = 0,$$

in terms of E considered as a variable, not an operator.

Example. To find y_m given that

$$y_{m+2} - \tfrac{3}{2} y_{m+1} + \tfrac{1}{8} y_m = 0 \text{ and } y_0 = 0, \ y_1 = \tfrac{1}{4}$$

We have

$$E^2 - \tfrac{3}{2}E + \tfrac{1}{8} = 0.$$

The roots of this quadratic in E are $E = x_1 = \tfrac{1}{2}$, $E = x_2 = \tfrac{1}{4}$. Hence

$$y_m = A(\tfrac{1}{2})^m + B(\tfrac{1}{4})^m.$$

Putting $m = 0$ we have

$$0 = A + B.$$

Putting $m = 1$ we have

$$\tfrac{1}{4} = \tfrac{1}{2}A + \tfrac{1}{4}B.$$

Hence

$$A = 1 \qquad B = -1$$

and

$$y_m = (\tfrac{1}{2})^m - (\tfrac{1}{4})^m.$$

Next we consider the case when some of the roots are not distinct, for example

$$y_{m+2} - 2a y_{m+1} + a^2 y_m = 0 \qquad a \neq 0$$

Hence

$$(E^2 - 2aE + a^2)\, y_m = (E - a)^2\, y_m = 0$$

so that

$$x_1 = x_2 = a.$$

and we have only one solution $y_m = Aa^m$ whereas we expect two independent solutions for a second order relation. Let the general solution be put in the form

$$y_m = u(m)\, a^m$$

where $u(m)$ is a function of m and not a constant. Then

$$(E - a)\, y_m = u(m + 1)\, a^{m+1} - au(m)\, a^m$$
$$= a^{m+1}\{u(m + 1) - u(m)\} = a^{m+1}\Delta u(m).$$

Similarly we can show that

$$(E - a)^2\, y_m = a^{m+2}\, \Delta^2 u(m),$$
$$(E - a)^3\, y_m = a^{m+3}\, \Delta^3 u(m),$$
$$\ldots \text{ etc.}$$

In our case we have

$$(E - a)^2\, y_m = a^{m+2}\, \Delta^2 u(m) = 0.$$

Hence

$$\Delta^2 u(m) = 0.$$

Hence $u(m)$ must be a polynomial of order 1, say $A + Bm$, and the required general solution is

$$y_m = (A + Bm)\, a^m$$

where A and B are arbitrary constants which can be determined from the starting values. In general, if $f(E) = 0$ has r roots equal to a then $(E - a)^r$ is a factor of $f(E)$ and the corresponding part of the general solution satisfies the relations

$$(E - a)^r\, y_m = 0,\ y_m = u(m)\, a^m,$$
$$\Delta^r u(m) = 0,$$

hence
$$u(m) = A_1 + A_2 m + A_3 m^2 + \ldots + A_r m^{r-1};$$
thus
$$y_m = (A_1 + A_2 m + A_3 m^2 + \ldots + A_r m^{r-1})\, a^m.$$

Next we consider recurrence relations with a non-zero right-hand side, for example of the form

$$y_{m+2} + a y_{m+1} + b y_m = g(m). \qquad (4.1.3)$$

First we consider the case when $g(m)$ is a polynomial, say of degree r. Since $E - 1 = \Delta$ and $\Delta^{r+1} g(m) = 0$, it follows that

$$(E - 1)^{r+1} (E^2 + aE + b)\, y_m = 0.$$

Hence the required solution is of the form

$$y_m = A x_1^m + B x_2^m + (A_0 + A_1 m + A_2 m^2 + \\ + \ldots + A_r m^{r-1}). \qquad (4.1.4)$$

In this solution A, B are determined from the starting values, but $(A_0, A_1, \ldots, A_r)$ are determined independently of the starting values by substituting (4.1.4) in (4.1.3) and equating coefficients. We may, however, proceed more directly as follows. Given any polynomial

$$F(x) = a_0 + a_1 x + a_2 x^2 + \ldots + a_n x^n, \qquad a_0 \neq 0,\ a_n \neq 0$$

we can find a polynomial

$$G(x) = b_0 + b_1 x + b_2 x^2 + \ldots + b_r x^r$$

such that

$$F(x)\, G(x) = 1 + \text{terms of order higher than } r \text{ in } x.$$

This is obtained by expanding the reciprocal of $F(x)$ as an infinite power series in x and then omitting all powers of x higher than r. In the case of (4.1.3),

$$E^2 + aE + b = (1 + \Delta)^2 + a(1 + \Delta) + b = F(\Delta),$$

and the required "particular" solution is

$$G(\Delta)\, g(m).$$

Example. To solve

$$y_{m+2} - \tfrac{3}{4}y_{m+1} + \tfrac{1}{8}y_m = m^2.$$

If we ignore the right-hand side we get, as before,

$$y_m = A(\tfrac{1}{2})^m + B(\tfrac{1}{4})^m.$$

To get the particular solution we have

$$(E^2 - \tfrac{3}{4}E + \tfrac{1}{8})\,y_m = [(1 + \Delta)^2 - \tfrac{3}{4}(1 + \Delta) + \tfrac{1}{8}]\,y_m$$
$$= \tfrac{3}{8}(1 + \tfrac{10}{3}\Delta + \tfrac{8}{3}\Delta^2)\,y_m = m^2.$$

We take
$$F(\Delta) = \tfrac{3}{8}(1 + \tfrac{10}{3}\Delta + \tfrac{8}{3}\Delta^2),$$
$$1/F(\Delta) = \tfrac{8}{3}(1 - \tfrac{10}{3}\Delta + \tfrac{76}{9}\Delta^2 - \dots).$$

We can ignore Δ^3 and higher powers of Δ as they annihilate m^2, hence
$$G(\Delta) = \tfrac{8}{3}\,(1 - \tfrac{10}{3}\Delta + \tfrac{76}{9}\Delta^2),$$
and
$$G(\Delta)\,m^2 = \tfrac{8}{3}\,[m^2 - \tfrac{10}{3}\,(2m + 1) + \tfrac{76}{9} \times 2]$$
$$= \tfrac{8}{3}\,(m^2 - \tfrac{20}{3}m + \tfrac{122}{9}).$$

The required general solution is

$$y_m = A(\tfrac{1}{2})^m + B\,(\tfrac{1}{4})^m + \tfrac{8}{3}\,(m^2 - \tfrac{20}{3}m + \tfrac{122}{9}).$$

If $F(\Delta)$ is easily factorized it may be easier to deal with the factors separately.

Thus in the last example

$$\frac{1}{F(\Delta)}\,m^2 = \tfrac{8}{3} \times \frac{1}{1 + \tfrac{4}{3}\Delta} \times \frac{1}{1 + 2\Delta}\,m^2,$$

$$= \tfrac{8}{3} \times \frac{1}{1 + \tfrac{4}{3}\Delta}\,(1 - 2\Delta + 4\Delta^2)\,m^2$$

$$= \tfrac{8}{3} \times \frac{1}{1 + \tfrac{4}{3}\Delta}\,(m^2 - 4m + 6),$$

$$= \tfrac{8}{3}\,(1 - \tfrac{4}{3}\Delta + \tfrac{16}{9}\Delta^2)(m^2 - 4m + 6)$$

$$= \tfrac{8}{3}\,(m^2 - \tfrac{20}{3}m + \tfrac{122}{9}).$$

Next we consider (4.1.3) when $g(m)$ is a power, e.g. $g(m) = t^m$. Since $F(\mathrm{E})\, t^m = F(t)\, t^m$, we see that provided $F(t) \neq 0$, the particular solution of

$$F(\mathrm{E})\, y_m = t^m$$

is

$$y_m = \frac{1}{F(t)}\, t^m.$$

If $F(t) = 0$ we can again put y_m in the form $u(m)\, t^m$ and derive $u(m)$ as in the case of equal roots.

Example. The general solution of

$$y_{m+2} - \tfrac{3}{4}y_{m+1} + \tfrac{1}{8}y_m = 2^m$$

is

$$y_m = A(\tfrac{1}{2})^m + B(\tfrac{1}{4})^m + \frac{1}{F}\, 2^m,$$

where

$$F = 2^2 - \tfrac{3}{4} \times 2 + \tfrac{1}{8} = \tfrac{21}{8},$$

and the general solution of

$$y_{m+2} - \tfrac{3}{4}y_{m+1} + \tfrac{1}{8}y_m = (\tfrac{1}{2})^m$$

is

$$y_m = A(\tfrac{1}{2})^m + B(\tfrac{1}{4})^m + 8m(\tfrac{1}{4})^m.$$

We now return to the solution of the recurrence relation

$$y_{m+2} + a y_{m+1} + b y_m = 0,$$

which we consider in order to illustrate the general case (4.1.2). The general solution is of the form

$$y_m = A(x_1)^m + B(x_2)^m,$$

where x_1, x_2 are the roots of the equation

$$\mathrm{E}^2 + a\mathrm{E} + b = 0.$$

If x_1, x_2 are complex and the coefficients a, b are real then we can put x_1, x_2 in the form

$$x_1 = re^{i\theta} \qquad x_2 = re^{-i\theta}$$

and the general solution can be expressed in the form

$$y_m = r^m \,(A \cos m\theta + B \sin m\theta) \ \text{ or } \ y_m = Ar^m \sin (m\theta + \alpha)$$

Even if x_1, x_2 are real it may be convenient to express the solution in terms of hyperbolic functions. Thus x_1, x_2 can be expressed in the form

$$x_1 = re^\theta, \ x_2 = re^{-\theta}, \ \text{where } r^2 = \pm b.$$

Then the general solution can be put in the form

$$y_m = r^m \,(A \cosh m\theta + B \sinh m\theta).$$

We conclude this section with some remarks concerning the use of recurrence relations in practical computations. Relations of the form (4.1.1):

$$y_{m+2} = ay_{m+1} + by_m$$

are frequently used in numerical analysis to compute consecutive y's because the coefficients a and b usually depend upon m and there is no explicit formula from which y_m can be computed easily so as not to involve the previous two y's. Nonetheless the analysis of this section still gives some idea of the behaviour of consecutive y's over a relatively short range of values of m, and sometimes even over a wide range. Thus it is important to study the numerical stability and the way in which error propagates in recurrence relations.

First we consider the case when the roots x_1, x_2 are distinct and

$$| x_1 | > | x_2 | .$$

Then in the formula for y_m:

$$y_m = A(x_1)^m + B(x_2)^m,$$

when m is sufficiently large the first term containing $(x_1)^m$ will be much larger than the term in $(x_2)^m$ and eventually the first term will dominate and "swamp" the second term if $A \neq 0$ so that

$$y_m \sim A(x_1)^m \text{ for } m \text{ sufficiently large.}$$

Any small round-off errors in the computation remain relatively small in the succeeding y's. The situation is different if $A = 0$ so that

$$y_m = B(x_2)^m$$

and the component of the larger root is absent. Rounding errors are bound to creep in when using (4.1.1) to compute consecutive y's. Such errors will usually introduce a component of the other larger root: $A(x_1)^m$. This error component, even though initially very small, grows rapidly relative to the true solution and soon the y's lose all accuracy. Sometimes this difficulty can be avoided by applying the recurrence relation (4.1.1) in reverse

$$y_m = -\frac{a}{b} y_{m+1} + \frac{1}{b} y_{m+2}.$$

Starting with given or suitably chosen starting values y_{N+1}, y_{N+2} we compute y_m for $m = N, N - 1, N - 2, \ldots, 1, 0$ in that order. The roots of this recurrence relation are $1/x_2$, $1/x_1$, but this time

$$\left| \frac{1}{x_2} \right| > \left| \frac{1}{x_1} \right|.$$

This instability does not occur if $|x_1| = |x_2|$, whether these roots are real or complex. Any small rounding errors introduced by applying (4.1.1) remain relatively small.

4.2 Summation of Series

(a) *Finite Summation*

Let

$$S(n) = y_1 + y_2 + \ldots + y_n$$

denote the sum of the first n terms of a series where the general term

$$y_r = y(r)$$

is some given function of r. The problem is to derive an explicit formula for $S(n)$. This can be done sometimes by establishing

4

some recurrence relation between $S(n)$, $S(n + 1)$, $S(n + 2)$, ..., and then using the methods of the previous section. Alternatively we can make use of the summation operator

$$\frac{1}{\Delta} = \frac{1}{E - 1}$$

introduced in §3.9. Thus

$$\begin{aligned}
S(n) &= y_1 + y_2 + \ldots + y_n \\
&= (1 + E + E^2 + \ldots + E^{n-1}) y_1, \\
&= \frac{E^n - 1}{E - 1} y_1.
\end{aligned} \tag{4.2.1}$$

Now $E^n - 1 = (1 + \Delta)^n - 1 = \binom{n}{1} \Delta + \binom{n}{2} \Delta^2 + \ldots$,

and $\quad \dfrac{E^n - 1}{E - 1} = \binom{n}{1} + \binom{n}{2} \Delta + \binom{n}{3} \Delta^2 + \ldots$,

hence $\quad S(n) = \binom{n}{1} y_1 + \binom{n}{2} \Delta y_1 + \binom{n}{3} \Delta^2 y_1 + \ldots$

Example. To find a formula for

$$S(n) = 1 + 4 + 9 + \ldots + n^2.$$

Here $\quad y(r) = r^2, \quad \Delta y(r) = 2r + 1, \quad \Delta^2 y(r) = 2.$

Third and higher differences of $y(r)$ are zero. Hence

$$S(n) = \binom{n}{1} + 3 \binom{n}{2} + 2 \binom{n}{3} = \frac{1}{6} n (n + 1)(2n + 1).$$

(b) Euler–Maclaurin Formula and Infinite Summation

If we expand $1/(e^x - 1)$ in ascending powers of x we get

$$\frac{1}{e^x - 1} = \frac{1}{x} - \tfrac{1}{2} + C_1 x + C_3 x^3 + C_5 x^5 + \ldots,$$

where $\quad C_{2r} = 0, \quad C_1 = \dfrac{1}{12}, \quad C_3 = \dfrac{-1}{720}, \quad C_5 = \dfrac{1}{30240}.$

If in this expression we replace x by hD and make use of the identity $E = e^{hD}$ established in §3.8, we get

$$\frac{1}{E - 1} = \frac{1}{hD} - \tfrac{1}{2} + C_1 hD + C_3 h^3 D^3 + \ldots \qquad (4.2.2)$$

Multiplying both sides by $E^n - 1$ and operating on y_0, the left-hand side becomes

$$\frac{E^n - 1}{E - 1} y_0 = y_0 + y_1 + \ldots + y_{n-1}.$$

In the right-hand side we interpret

$$(E^n - 1)\frac{1}{hD} y_0 \text{ as } \frac{1}{h} \int_{x_0}^{x_n} y(x)\,dx$$

where $y_r = y(x_0 + rh)$. Then (4.2.2) becomes the Euler–Maclaurin formula:

$$\tfrac{1}{2}y_0 + y_1 + y_2 + \ldots + y_{n-1} + \tfrac{1}{2}y_n$$

$$= \frac{1}{h} \int_{x_0}^{x_n} y(x)\,dx + C_1 h(y_n' - y_0') + C_3 h^3(y_n''' - y_0''') +$$

$$+ \ldots + C_{2r-1} h^{2r-1}(y_n^{(2r-1)} - y_0^{(2r-1)}) + R,$$

where R is the remainder term of the type discussed in §3.9. It can be shown that

$$R = n C_{2r+1} h^{2r+1} y^{(2r)}(u), \text{ where } x_0 \leqq u \leqq x_n.$$

The Euler–Maclaurin formula can be considered as a formula for evaluating an integral over several intervals like (3.9.5); it is more efficient because the coefficients are smaller but it assumes a knowledge of the derivatives at the two ends. It leads to a useful result when n tends to infinity. Suppose all of the following conditions are satisfied:

(i) the infinite series $y_0 + y_1 + y_2 + \ldots$ is convergent,

(ii) the infinite integral $\int\limits_{x_0}^{\infty} y(x)\, \mathrm{d}x$ is convergent,

(iii) the infinite series $C_1 y_0' + C_3 y_0''' + \ldots$ is convergent,

(iv) not only y_n but also all its derivatives y_n', y_n'', ..., tend to zero when n tends to infinity,

(v) the remainder R tends to zero as r tends to infinity, i.e. $y^{(2r)}(x)$ is negligible throughout the range $x > x_0$ if the series in (iii) is terminated after r terms.

Then, when n tends to infinity, the Euler–Maclaurin formula becomes

$$y_0 + y_1 + y_2 + \ldots = \frac{1}{h} \int\limits_{x_0}^{\infty} y(x)\, \mathrm{d}x + \tfrac{1}{2}y_0 -$$
$$- hC_1 y_0' - h^3 C_3 y_0''' - \ldots \ldots (4.2.3)$$

This result can be considered as a formula for evaluating the sum of the infinite series on the left if the infinite integral on the right is known and if, as often happens, the infinite series on the right is rapidly convergent. Alternatively, (4.2.3) can be considered as a formula for evaluating an infinite integral if the series on the left and right are easy to sum.

Example. To sum the series

$$1/100^2 + 1/101^2 + 1/102^2 + \ldots ,$$

which is very slowly convergent, take

$$y(x) = 1/x^2, \qquad x_0 = 100, \qquad h = 1,$$

so that $\qquad\qquad y_0 = 0 \cdot 0001,$

$$y'(x) = -2/x^3, \qquad y_0' = -0 \cdot 000002,$$

$$\int\limits_{y_0}^{\infty} y(x)\, \mathrm{d}x = 0 \cdot 01.$$

Substituting these values of the first three terms in the right-hand side of (4.2.3) we get $0 \cdot 0100501667$, which is correct to ten decimal places. To get the same accuracy by summing directly the given series would require a number of terms of the order of thousands.

(c) *Summation of Alternating Series*

Let
$$S = y_0 - y_1 + y_2 - y_3 + \ldots ,$$

$$S_n = y_0 - y_1 + y_2 - \ldots \pm y_n$$

$$= (1 - E + E^2 - \ldots \pm E^n)\, y_0 = \frac{1 \pm E^{n+1}}{1 + E}\, y_0$$

$$= \frac{1}{1 + E}\, y_0 \pm \frac{1}{1 + E}\, y_{n+1}.$$

If the series is convergent the second term tends to zero and we get

$$S = \frac{1}{1 + E}\, y_0 = \frac{1}{2 + \Delta}\, y_0 = \tfrac{1}{2}(1 - \tfrac{1}{2}\Delta + \tfrac{1}{4}\Delta^2 - \ldots)\, y_0$$

$$= \tfrac{1}{2}(y_0 - \tfrac{1}{2}\Delta y_0 + \tfrac{1}{4}\Delta^2 y_0 - \tfrac{1}{8}\Delta^3 y_0 + \ldots).$$

This series is often more rapidly convergent than the given one.

Example. To sum the series
$$1 - \tfrac{1}{2} + \tfrac{1}{3} - \tfrac{1}{4} + \tfrac{1}{5} - \ldots$$

we put it in the form
$$1 - \tfrac{1}{2} + \tfrac{1}{3} - S = 0 \cdot 8333 - S,$$
where
$$S = \tfrac{1}{4} - \tfrac{1}{5} + \tfrac{1}{6} - \ldots = y_0 - y_1 + y_2 - \ldots$$

We construct the difference table:

$$
\begin{array}{cccccc}
0\cdot2500 \\
& -500 \\
0\cdot2000 & & 167 \\
& -333 & & -72 \\
0\cdot1667 & & 95 & & 36 \\
& -238 & & -36 \\
0\cdot1429 & & 59 \\
& -179 \\
0\cdot1250 \\
\end{array}
$$

Hence

$$
\begin{aligned}
S &= \tfrac{1}{2}(0\cdot2500 - \tfrac{1}{2} \times - 0\cdot0500 + \tfrac{1}{4} \times 0\cdot0167 - \\
& \qquad\qquad - \tfrac{1}{8} \times 0\cdot0072 + \tfrac{1}{16} \times 0\cdot0036) \\
&= 0\cdot1401,
\end{aligned}
$$

and the required sum is $0\cdot8333 - 0\cdot1401 = 0\cdot6932$ which is ln 2 correct to 4 decimal places. To get the same accuracy by summing directly the given series would require a number of terms of the order of thousands.

4.3 Bernoulli's Method for Solving Polynomial Equations

It was explained in §4.1 that if

$$(x_1, x_2, \ldots, x_n)$$

are the roots of the equation

$$x^n + a_1x^{n-1} + a_2x^{n-2} + \ldots + a_n = 0,$$

then the general solution of the recurrence relation

$$y_{m+n} + a_1y_{m+n-1} + a_2y_{m+n-2} + \ldots + a_my_m = 0$$

is of the form

$$y_m = A_1(x_1)^m + A_2(x_2)^m + \ldots + A_n(x_n)^m. \qquad (4.3.1)$$

This result could be used in the reverse order to find the roots of a polynomial. In the simplest case if

$$|x_1| > |x_2| > \ldots > |x_n|$$

then $A_1(x_1)^m$ is much the largest term in (4.3.1). For example, if

$$|x_2| = 0 \cdot 5 |x_1|,$$

then

$$|x_2^{10}| \sim 0 \cdot 001 |x_1^{10}|.$$

Thus although x_2 may be an appreciable fraction of x_1, x_2^m is a negligible fraction of x_1^m when m is sufficiently large. Then

$$y_m \sim A_1(x_1)^m$$

and

$$x_1 \sim y_{m+1}/y_m$$

The largest root is the limit of the ratio of two consecutive y's. This method for finding the numerically largest root is called Bernoulli's method.

Example 1. To find the numerically largest root of the equation

$$x^2 - x - 1 = 0$$

we consider the recurrence relation

$$y_{m+2} - y_{m+1} - y_m = 0.$$

Starting with two arbitrary values of y_0 and y_1, say $y_0 = 1$, $y_1 = 1$, and using $y_{m+2} = y_{m+1} + y_m$ we get

m	y_m	y_m/y_{m-1}
0	1	
1	1	1
2	2	2
3	3	1·5
4	5	1·67
5	8	1·60
6	13	1·625
7	21	1·616
8	34	1·619
9	55	1·6177
10	89	1·61818

The correct answer is $\frac{1}{2}(1 + \sqrt{5}) = 1 \cdot 61803$.

Next we consider the case when the two largest roots have the same modulus, as happens when they are complex conjugates. Then, by an argument similar to that in the previous case, we have when m is sufficiently large

$$y_m \sim A_1 x_1^m + A_2 x_2^m.$$

If x_1, x_2 are the roots of the quadratic equation

$$x^2 + px + q = 0$$

it follows that

$$
\begin{aligned}
y_{m+2} &+ py_{m+1} + qy_m \\
&\sim (A_1 x_1^{m+2} + A_2 x_2^{m+2}) + p(A_1 x_1^{m+1} + A_2 x_2^{m+1}) + \\
&\qquad\qquad\qquad\qquad\qquad\quad + q(A_1 x_1^m + A_2 x_2^m) \\
&= A_1 x_1^m (x_1^2 + px_1 + q) + A_2 x_2^m (x_2^2 + px_2 + q) \\
&= 0
\end{aligned}
$$

Thus
$$y_{m+2} + py_{m+1} + qy_m \sim 0$$

and similarly
$$y_{m+3} + py_{m+2} + qy_{m+1} \sim 0.$$

We solve these two as simultaneous exact equations in p and q for consecutive values of m and stop when p and q settle to the desired degree of accuracy.

Example 2. To find the largest pair of roots of the equation

$$x^4 - x^3 + 6x^2 + 5x + 10 = 0.$$

In this case the largest two roots are complex conjugate and so are the remaining two. To find the roots it is sufficient to find the corresponding quadratic factor

$$x^2 + px + q.$$

We take arbitrary starting values

$$y_0 = 4, \qquad y_1 = 1, \qquad y_2 = -11, \qquad y_3 = -32.$$

Then from the recurrence relation

$$y_{m+4} - y_{m+3} + 6y_{m+2} + 5y_{m+1} + 10y_m = 0$$

we have

m	y_m	p	q
4	−11		
5	226		
6	562		
7	−419		
8	−4811		
9	−7367		
10	17974		
11	90421		
12	67522	−2·146	7·038
13	−491204	−2·145	7·033

We do not calculate p and q for low values of m as it is unlikely that they will have much accuracy then. Thus the largest pair of roots are the roots of

$$x^2 - 2\cdot145x + 7\cdot033 = 0.$$

This method can be used when the two largest roots have nearly the same modulus when the direct method for finding the largest root converges very slowly. It can also be used to find the second largest root after having found the largest root. For example, having found x_1 as the ratio y_{20}/y_{19}, to find x_2 we compute p, q so that

$$y_{20} + py_{19} + qy_{18} = 0,$$
$$y_{19} + py_{18} + qy_{17} = 0,$$

and x_2 is the second root of the quadratic equation

$$x^2 + px + q = 0.$$

This method can be extended in an obvious manner to find the 3 or 4 largest roots if they all happen to have the same or nearly the same modulus.

Bernoulli's method for finding the largest root, and its extension for finding the larger roots, of a polynomial are frequently used in numerical analysis in many contexts besides the problem of polynomial equations. They are easy to program, but convergence is usually slow. When we are approximating to the largest root x_1, the rate of convergence depends upon its separation from x_2, i.e. upon the ratio x_2/x_1; as was explained at the beginning of this section, the smaller the ratio the faster is the convergence.

There are various techniques for speeding up the convergence. Only the simplest two methods will be explained here. The first is by shifting the origin: given a polynomial equation $f(x) = 0$ it is easy to derive a new equation $g(x) = 0$ whose roots are those of $f(x)$ shifted by a constant k. This is done by replacing x by $x + k$ in $f(x)$. Thus if $x_1 = 105$ and $x_2 = 101$ then $x_2/x_1 \sim 1$; but if we choose $k = 100$ then $(x_2 - k)/(x_1 - k) = 0 \cdot 2$ which is a much better separation and leads to much faster convergence. In practice this is not always possible as a large shift will upset the numerical order of magnitudes of the roots so that the dominant root x_1 may become subdominant.

The second is Aitken's accelerating formula, which is applicable to any iterative procedure. A sequence of numbers $u_1, u_2, \ldots, u_n, u_{n+1}, u_{n+2}, \ldots$, is expected to converge to an unknown limit u, but is converging slowly. In order to find a good estimate of u we assume that

$$u_n = u - e \qquad u_{n+1} = u - ke \qquad u_{n+2} = u - k^2 e \ldots$$

so that the error is tending to zero in a geometric progression. It is first e, then ke, then $k^2 e$, $\ldots$, where the common ratio k is numerically less than, but nearly equal to, 1. By eliminating k and e from the three equations we get

$$u = \frac{u_{n+2} u_n - u_{n+1}^2}{u_{n+2} - 2u_{n+1} + u_n} = u_{n+2} - \frac{(u_{n+2} - u_{n+1})^2}{u_{n+2} - 2u_{n+1} + u_n}$$

As was explained in Exercise 9 of Chapter 2, the second expression is better for numerical evaluation because the numerator is usually much smaller than the denominator which is itself small owing to a lot of cancellation.

Example 3. In Example 1 we have

$$y_8 = 1 \cdot 619048$$

$$y_9 = 1 \cdot 617647$$

$$y_{10} = 1 \cdot 618182$$

$$y = 1 \cdot 618034 \text{ (using Aitken's formula)}$$

Thus y has six decimal places correct whereas y_{10} has only three decimal places correct.

Bernoulli's method can be used to find the smallest root of a polynomial equation by applying the recurrence relation in reverse. By a suitable shift of origin, it can be used to find intermediate roots. The various techniques described in this section are not as good as the methods described in the next sections for finding the roots of a polynomial; but they are suitable for finding an initial rough approximation. They are also important because they apply to problems which arise in matrices and other iterative processes.

4.4 Newton–Raphson Method. Order of Convergence of an Iterative Procedure

To find a root x of an algebraic or transcendental equation $f(x) = 0$ we start with a given approximation x_n and seek an improved approximation x_{n+1}. Let e_n, e_{n+1} be the respective errors in x_n, x_{n+1} so that

$$x_n = x + e_n,$$

$$x_{n+1} = x + e_{n+1}.$$

For brevity we write

$$f, f', f_n, f_n', \ldots, \text{ for } f(x), f'(x), f(x_n), f'(x_n), \ldots$$

Expanding by Taylor's series we get

$$0 = f(x) = f(x_n - e_n) = f_n - e_n f_n' + \tfrac{1}{2} e_n^2 f_n'' - \ldots$$

If $f_n' \neq 0$ and if we ignore e_n^2 and higher powers we get

$$e_n \sim f_n/f_n',$$

$$x = x_n - e_n \sim x_n - f_n/f_n'.$$

An improved approximation to x is then

$$x_{n+1} = x_n - f_n/f_n',$$

which is the Newton–Raphson formula.

Example. To find the root of the equation $x^2 - x - 1 = 0$ which is nearest to 2. We have

$$f(x) = x^2 - x - 1, \qquad f'(x) = 2x - 1,$$

$$x_0 = 2, \qquad x_1 = x_0 - f_0/f_0' = 2 - \tfrac{1}{3} = 1 \cdot 667,$$

$$x_1 = 1 \cdot 667, \qquad x_2 = x_1 - f_1/f_1' = 1 \cdot 667 - 0 \cdot 1111/2 \cdot 333$$

$$= 1 \cdot 619.$$

Comparing this result with that of Example 1 of the previous section, the convergence is remarkably rapid. It is worthwhile studying what determines the rate of convergence of various iterative processes. We begin with the Newton–Raphson method.

Newton's formula expressed in terms of e_n, e_{n+1} becomes

$$x + e_{n+1} = x + e_n - f_n/f_n',$$

hence $\qquad\qquad e_{n+1}f_n' = e_n f_n' - f_n.$

By Taylor's series we have

$$f_n = f(x + e_n) = e_n f' + \tfrac{1}{2}e_n^2 f'' + \dots,$$

$$f_n' = f'(x + e_n) = f' + e_n f'' + \dots.$$

Substituting these values we get

$$e_{n+1}(f' + e_n f'' + \dots) = e_n(f' + e_n f'' + \dots) -$$
$$- (e_n f' + \tfrac{1}{2}e_n^2 f'' + \dots).$$

Hence, provided $f' \neq 0$, e_{n+1} is of order e_n^2; neglecting terms of higher order we get

$$e_{n+1} = ke_n^2, \text{ where } k = \tfrac{1}{2}f''/f'.$$

Next we consider Bernoulli's method. Taking the case of Example 1 in the previous section we have

$$y_n = Ax_1^n + Bx_2^n = Ax_1^n \left(1 + \frac{B}{A}k^n\right), \text{ where } k = x_2/x_1,$$

hence $$y_{n+1}/y_n = x_1 + e_n$$

where, ignoring terms of higher order,

$$e_n = \frac{B}{A}(k-1)k^n x_1$$

so $$e_{n+1} = ke_n.$$

We have now two formulae for the rate of convergence:

$$e_{n+1} = ke_n, \qquad \text{(Bernoulli's method)} \qquad (4.4.1)$$

$$e_{n+1} = ke_n^2. \qquad \text{(Newton–Raphson method)} \qquad (4.4.2)$$

The first is said to be *linear* or *first order* convergence; for convergence k has to be numerically less than 1, the smaller it is the faster the convergence; this type of convergence is typical of many matrix and other iterative processes where it often happens, unfortunately, that k is very nearly equal to 1 and convergence is therefore very slow. In the second formula convergence is said to be *quadratic* or *second order*; it is far more rapid than linear convergence and k need not be numerically less than 1. Similarly there are iterative processes for which the convergence is of order 3 (i.e. $e_{n+1} = ke_n^3$) or higher. The higher the order the more rapid is the convergence; but unfortunately, in order to obtain convergence formulae of order higher than 2 one usually has to pay the penalty of more complicated formulae and calculations. When assessing the efficiency of an iterative procedure one must take into account the amount of computing per iteration as well as the order of convergence.

In the rest of this section we consider the case when $f(x)$ is a polynomial

$$f(x) = x^n + a_1 x^{n-1} + a_2 x^{n-2} + \ldots + a_n. \quad (4.4.3)$$

To evaluate $f(x)$ and $f'(x)$ for a given approximation p we use the method of synthetic division. Let

$$f(x) = (x - p)\, Q(x) + b_n$$

where $\qquad Q(x) = x^{n-1} + b_1 x^{n-2} + \ldots + b_{n-1} \quad (4.4.4)$

$Q(x)$, b_n are respectively the quotient and remainder when $f(x)$ is divided by $(x - p)$. Then

$$f(p) = b_n, \qquad f'(p) = Q(p).$$

These two quantities can be computed by applying simultaneously the recurrence relations

$$b_r = a_r + p b_{r-1}, \qquad b_0 = 1 \qquad c_0 = 0$$
$$c_r = b_{r-1} + p c_{r-1}, \qquad r = 1, 2, \ldots, n.$$

The first recurrence relation is derived by equating coefficients of equal powers of x in (4.4.4). The second recurrence relation is derived by evaluating the polynomial $Q(x)$ by the "box" method explained in §§1.6 and 2.2; the first recurrence relation may also be considered as a method of evaluating $f(x)$ by the same technique. The required improved approximation to the root of $f(x)$ is

$$p - f(p)/f'(p) = p - b_n/c_n.$$

When $f(p)$ is sufficiently small p is taken as the required root. To find the next root of $f(x)$ we replace it by

$$Q(x) = f(x)/(x - p) = x^{n-1} + b_1 x^{n-2} + \ldots + b_{n-1}$$

This process is called *deflation*. No extra computation is required to find the coefficients of the new reduced polynomial as these

coefficients
$$(1, b_1, b_2, \ldots, b_{n-1})$$

are a by-product of the Newton–Raphson procedure for the previous root. Thus the degree of the polynomial is successively reduced after each root extraction. After finding a root of the deflated polynomial we could improve accuracy by applying the Newton–Raphson procedure to the original polynomial.

A polynomial is said to be *ill-conditioned* if a relatively small change in the coefficients causes a relatively large change in the roots. For such polynomials the result of applying the Newton–Raphson (or any other) method may give roots with very few or even no correct significant figures. Such polynomials may occur frequently if the degree n is large or if some of the roots are equal or nearly equal. This warning is especially necessary for automatic programs.

The choice of a suitable initial approximation in an automatic program is important but not easy. Generally it is safe to find the numerically smallest root first because the deflation procedure is then numerically more stable. In this connexion the following inequality may be of some help. If x is a root of $f(x)$, then

$$| a_n | /R < | x | < R,$$

where

$$R = 1 + | a_1 | + | a_2 | + \ldots + | a_n | .$$

If the coefficients $(a_1, a_2, \ldots, a_n)$ are real and all the roots are known to be real, the Newton–Raphson and deflation procedures can be carried out using ordinary real arithmetic operations. If some of the roots are complex, as is usually the case, then we can still apply the above procedures using complex arithmetic however; this is also done if the coefficients are complex. For real coefficients one can speed up the computation by making use of the fact that complex roots occur in conjugate pairs. In such cases the initial approximation must be complex and not real or purely imaginary, otherwise all intermediate results and successive approximations may remain real and complex roots remain undetected.

4.5 Graeffe's Root–Squaring Method

The method considered in this section applies in practice to
polynomial equations only. It is usually not as efficient as the
methods of the previous section or the next section. But it is
worth studying because it illustrates a useful principle which can
be used in other contexts. Consider, for example, the quartic
equation
$$x^4 - ax^3 + bx^2 - cx + d = 0. \qquad (4.5.1)$$

Separating even and odd powers of x on the right and left of the
equation we get
$$x^4 + bx^2 + d = ax^3 + cx.$$

Squaring and replacing x^2 by X we get
$$X^4 - a'X^3 + b'X^2 - c'X + d' = 0$$

where
$$\left.\begin{aligned}
a' &= a^2 - 2b, \\
b' &= b^2 - 2ac + 2d, \\
c' &= c^2 - 2bd, \\
d' &= d^2.
\end{aligned}\right\} \qquad (4.5.2)$$

Thus the roots of
$$x^4 - a'x^3 + b'x^2 - c'x + d' = 0 \qquad (4.5.3)[1]$$

are the squares of the roots of (4.5.1). If the transformation
(4.5.2) is applied n times we get an equation $(4.5.3)^n$ whose
roots are
$$x_1^{2^n}, \qquad x_2^{2^n}, \qquad x_3^{2^n}, \qquad x_4^{2^n}$$

where x_1, x_2, x_3, x_4 are the roots of the original equation (4.5.1).
We arrange the roots in order of numerical magnitude so that
$$|x_1| > |x_2| \geqq |x_3| \geqq |x_4|$$

If the first root is well separated than $x_1^{2^n}$ will soon dominate all
the others. From equation $(4.5.3)^n$ we have

$$a' = \text{sum of the roots} = x_1^{2^n} + x_2^{2^n} + x_3^{2^n} + x_4^{2^n} \sim x_1^{2^n},$$

hence
$$x_1 = \pm (a')^{1/2^n}.$$

We do not consider the complex $2n$th roots of a' because x_1 must be real if it is separated from x_2; the sign of x_1 can be determined by substituting in the given equation. This method is rapidly convergent; in fact it can be shown that convergence is of second order, but unfortunately it cannot be applied quite so easily when the largest pair of roots are of equal moduli, which is what happens when x_1 and x_2 are complex conjugate. In such a case a' will oscillate and there will be no convergence. However, in $(4.5.3)^n$ we have

$$b' = \text{sum of the products of the roots two at a time}$$
$$= (x_1 x_2)^{2^n} + (x_1 x_3)^{2^n} + \ldots \sim (x_1 x_2)^{2^n}$$

if $\qquad |x_1| = |x_2| > |x_3| \geqq |x_4|.$

Although $(a')^{1/2^n}$ oscillates, $+(b')^{1/2^n}$ will converge to $x_1 x_2$ which must be positive when x_1, x_2 are complex conjugates.

Example. To find the product $x_1 x_2$ for the equation

$$x^4 - x^3 + 6x^2 + 5x + 10 = 0$$

of Example 2 in §4.3. Applying formulae $(4.5.2)$ we get

n	1	2	3	4
a'	-11	-11	-4811	$1 \cdot 11271 \times 10^7$
b'	66	2466	$6 \cdot 00931 \times 10^7$	$3 \cdot 58052 \times 10^{13}$
c'	-95	-4175	$-3 \cdot 18894 \times 10^7$	$-1 \cdot 84928 \times 10^{14}$
d'	10^2	10^4	10^8	10^{16}

It will be found that the next iteration will simply square b' and no further useful convergence is obtained if the arithmetic is carried to 6 significant figures. Thus

$$(x_1 x_2)^{16} = 3 \cdot 58052 \times 10^{13}.$$

Hence, using six figure tables,

$$x_1 x_2 = 7 \cdot 03269.$$

Unfortunately, this is not enough to determine separately the complex conjugate roots x_1 and x_2, and a certain amount of awkward calculation is necessary in general. But for the case of an equation of order 4 it is possible to derive the largest pair of roots by some elementary calculations. In the above example let

$$x^4 - x^3 + 6x^2 + 5x + 10 = (x^2 + px + q)(x^2 + Px + Q)$$

where P, Q refer to the larger pair of roots, so that

$$Q = x_1 x_2 = 7 \cdot 03269.$$

Equating constant terms we have

$$Qq = 10 \text{ hence } q = 1 \cdot 42193.$$

Equating coefficients of x^3 and x we get

$$P + p = -1$$

$$1 \cdot 42193P + 7 \cdot 03269p = 5,$$

hence $P = -2 \cdot 14458, \qquad p = 1 \cdot 14458.$

The final answer is

$$x^4 - x^3 + 6x^2 + 5x + 10$$
$$= (x^2 - 2 \cdot 14458x + 7 \cdot 03269)(x^2 + 1 \cdot 14458x + 1 \cdot 42193).$$

Check: coefficient of $x^2 = 5 \cdot 99998.$

4.6 The Methods of Lin and Bairstow

When the largest pair of roots of a polynomial are complex conjugate, they have the same modulus; the methods of Bernoulli and Graeffe are no longer so straightforward to apply; the Newton–Raphson method could be applied but this would require complex arithmetic. The object of this section is to find in conjugate pairs the complex roots of a polynomial with real coefficients. Thus, given a polynomial

$$f(x) = x^n + a_1 x^{n-1} + a_2 x^{n-2} + \ldots + a_n$$

and an approximate quadratic factor

$$H(x) = x^2 + px + q,$$

we seek an improved approximation to this quadratic factor

$$x^2 + p'x + q',$$

where all the coefficients $a_1, a_2, \ldots, a_n, p, q, p', q'$ are real and thus all the arithmetic operations will be real. The complex conjugate roots of the improved factor are the required roots of the given polynomial. Let

$$f(x) = H(x)\,Q_1(x) + R_1(x) \qquad (4.6.1)$$

where

$$Q_1(x) = x^{n-2} + b_1 x^{n-3} + b_2 x^{n-4} + \ldots + b_{n-2},$$

$$R_1(x) = b_{n-1}(x + p) + b_n.$$

Thus $Q_1(x)$ is the quotient and $R_1(x)$ is the remainder when $f(x)$ is divided by the quadratic factor $H(x)$. The remainder $R_1(x)$ is expressed in the form given because we can establish a simple recurrence relation for computing the coefficients b_r for $r = 1, 2, \ldots, n-1, n$. By equating the coefficients of x^r on the right and left of (4.6.1) we get

$$a_r = b_r + pb_{r-1} + qb_{r-2} \text{ for } r = 1, 2, \ldots, n, \qquad (4.6.2)$$

provided we take $b_{-1} = 0$, $b_0 = 1$. Hence given these two starting values we can compute b_r by solving (4.6.2) in the form

$$b_r = a_r - pb_{r-1} - qb_{r-2}, \qquad \left.\begin{array}{l} b_{-1} = 0, \quad b_0 = 1, \\ r = 1, 2, \ldots, n. \end{array}\right\} \quad (4.6.3)$$

This method of carrying out synthetic division by a quadratic factor is an extension of the method of division by a linear factor explained in §4.4; the recurrence relations are easy to program on a computer.

For the derivation of the methods of Lin and Bairstow it would help to think of the coefficients b_r as functions in p and q,

$$b_r = b_r(p, q),$$

as their values depend upon p and q which will vary, and upon the coefficients a_1, a_2, ... , a_n which are fixed. The problem is then to determine p and q so that the remainder is zero, i.e.

$$\left.\begin{aligned}
b_{n-1} = b_{n-1}(p, q) = a_{n-1} - pb_{n-2} - qb_{n-3} = 0, \\
b_n = b_n(p, q) = a_n - pb_{n-1} - qb_{n-2} = 0.
\end{aligned}\right\} \quad (4.6.4)$$

Lin's method is to solve these two equations for (p', q') by replacing b_n and b_{n-1} by zero and (p, q) by (p', q'):

$$\left.\begin{aligned}
a_{n-1} - p'b_{n-2} - q'b_{n-3} = 0, \\
a_n - q'b_{n-2} = 0.
\end{aligned}\right\} \quad (4.6.5)$$

Example 1. Given that

$$x^2 + 1 \cdot 14x + 1 \cdot 42$$

is an approximate factor of

$$x^4 - x^3 + 6x^2 + 5x + 10;$$

(see Example 2, §4.3) to improve the approximation we use (4.6.5) repeatedly, giving

	First	Second	Third	Fourth	Fifth
p	1·14	1·146590	1·144087	1·144653	1·144567
q	1·42	1·424582	1·421125	1·422091	1·421911
b_1	−2·14	−2·156590	−2·144087	−2·144653	−2·144567
b_2	7·0196	7·036677	7·031897	7·032792	7·032690
b_3	0·036456	−0·010190	0·001914	−0·000315	−0·000002
b_4	−0·009392	−0·012640	0·004606	−0·001074	0·000143
q'	1·424582	1·421125	1·422091	1·421911	1·421931
p'	1·146590	1·144087	1·144653	1·144567	1·144572

The required factor is

$$x^2 + 1 \cdot 144572x + 1 \cdot 421931.$$

It can be shown that the convergence of Lin's method is of order 1, i.e. the errors in consecutive approximations obey a law of the form

$$e_{r+1} = ke_r,$$

and the procedure is convergent only if k is numerically less than 1. Thus Lin's method may not converge, and when it does convergence is usually slow.

Bairstow's method, to be described next, has second order convergence; it is therefore much faster. In fact it is Newton's method applied to the simultaneous equations (4.6.4) in the two unknowns (p, q). Let

$$Q_1(x) = H(x) Q_2(x) + R_2(x), \qquad (4.6.6)$$

where

$$Q_2(x) = x^{n-4} + c_1 x^{n-5} + \ldots + c_{n-4},$$

$$R_2(x) = c_{n-3}(x + p) + c_{n-2}$$

are respectively the quotient and remainder when $Q_1(x)$ is divided by $H(x)$ so that

$$f(x) = H^2(x) Q_2(x) + H(x) R_2(x) + R_1(x).$$

By equating the coefficients of x^r in (4.6.6) we get the relation similar to (4.6.2)

$$b_r = c_r + pc_{r-1} + qc_{r-2} \qquad (4.6.7)$$

and hence we can compute c_r from the recurrence relation

$$\left. \begin{array}{l} c_0 = 1, \ c_{-1} = 0, \\ c_r = b_r - pc_{r-1} - qc_{r-2}, \quad r = 1, 2, \ldots, n-2, n-1. \end{array} \right\} \quad (4.6.8)$$

Let

$$A_r = \frac{\partial b_r}{\partial p}, \qquad B_r = \frac{\partial b_r}{\partial q}.$$

By differentiating (4.6.2) with respect to p we get

$$0 = A_r + pA_{r-1} + b_{r-1} + qA_{r-2},$$

hence

$$-b_{r-1} = A_r + pA_{r-1} + qA_{r-2}.$$

Similarly by differentiating (4.6.2) with respect to q we get

$$- b_{r-2} = B_r + pB_{r-1} + qB_{r-2}.$$

Thus we see that A_r and B_r satisfy the same recurrence relation as $-c_{r-1}$ and $-c_{r-2}$ respectively and we can also verify that they have the same starting values, hence

$$\left. \begin{aligned} A_r &= - c_{r-1}, \\ B_r &= - c_{r-2}. \end{aligned} \right\} \tag{4.6.9}$$

Let

$$\left. \begin{aligned} p' &= p + \delta p, \\ q' &= q + \delta q \end{aligned} \right\} \tag{4.6.10}$$

be the required improved approximation to (p, q) so that the x coefficient and the constant in R_1 (4.6.1.) are zero, i.e.

$$b_{n-1}(p', q') = 0,$$

$$p' b_{n-1}(p', q') + b_n(p', q') = 0.$$

Now

$$b_n(p', q') = b_n(p + \delta p, \; q + \delta q)$$

$$\sim b_n(p, q) + A_n \delta p + B_n \delta q$$

$$= b_n - c_{n-1}\delta p - c_{n-2}\delta q$$

and similarly

$$b_{n-1}(p', q') \sim b_{n-1} - c_{n-2}\delta p - c_{n-3}\delta q.$$

Hence, ignoring terms of higher order, $(\delta p, \delta q)$ are the solution of the simultaneous equations

$$\left. \begin{aligned} (c_{n-1} - b_{n-1})\delta p + c_{n-2}\delta q &= b_n, \\ c_{n-2}\delta p + c_{n-3}\delta q &= b_{n-1} \end{aligned} \right\} \tag{4.6.11}$$

To sum up, Bairstow's method consists of applying the recurrence relations (4.6.3) and (4.6.8) simultaneously or consecutively to compute $b_1, \ldots, b_n$ and $c_1, \ldots, c_{n-1}$; then using (4.6.11) to compute δp and δq, and (4.6.10) to compute (p', q').

Example 2. Taking the same initial quadratic factor as in Example 1 we have:

p	1·14	1·144583	1·144574
q	1·42	1·421933	1·421931
b_1	−2·14	−2·144583	−2·144574
b_2	7·0196	7·032720	7·032692
b_3	0·036456	−0·000083	−0·000000
b_4	−0·009392	0·000033	0·000000
c_1	−3·28	−3·289176	
c_2	9·3388	9·375522	
c_3	−5·952176	−6·054153	
δp	0·004583	−0·000009	
δq	0·001933	−0·000002	

The coefficients b_1, b_2 give the second quadratic factor, and the final answer is

$$x^4 - x^3 + 6x^2 + 5x + 10$$
$$= (x^2 + 1·144574x + 1·421931) \times$$
$$\times (x^2 - 2·144574x + 7·032692).$$

The same remarks concerning deflation in the Newton–Raphson process apply to Bairstow's method. After extracting a quadratic factor the degree of the polynomial is reduced by 2 and $f(x)$ is replaced by $Q_1(x)$ which is a by-product of the Bairstow process. Bairstow's method is also suitable for finding a pair of nearly equal real roots. The main disadvantage of this method is that it requires an initial approximation to the required quadratic factor. Such an initial approximation could be derived by applying the extension of Bernoulli's method explained in §4.3.

We conclude this chapter with some brief remarks concerning further methods which are not treated in this book; for further details consult the references mentioned in the suggestions for further reading. When $f(x)$ is not a polynomial the Newton–Raphson method is still efficient provided $f'(x)$ is easy to evaluate; otherwise more powerful methods have been developed where we do not require the evaluation of the derivative. The modified

Newton method replaces $f'(x)$ by the gradient of the chord joining the latest two points, i.e. it fits a straight line through the latest two approximations. Muller's method fits a parabola through the latest three approximations. Another important subject is that of finding the zeros of simultaneous algebraic or transcendental equations; the Newton–Raphson method can be extended to this case, this was briefly illustrated in the derivation of Bairstow's method.

EXERCISES

1. Solve the difference equation:

$$y_{n+2} + 6y_{n+1} - 7y_n = 3^n.$$ (U.L.)

2. Show that the equation:

$$y_{n+1}y_n + 2y_{n+1} + 8y_n + 9 = 0$$

can be reduced by the substitution

$$y_n = V_{n+1}/V_n - 2$$

to an equation with constant coefficients. Hence solve the given equation when $y_0 = -5$. (U.L.)

3. If

$$y_{n+2} - y_{n+1} + y_n = 0, \qquad y_1 = 0, \qquad y_2 = 1,$$

show that

$$y_n = \frac{\sin\left[(n-1)\,\pi/3\right]}{\sin\left(\pi/3\right)}.$$ (U.L.)

4. If

$$y_{n+2} + 4y_{n+1} + y_n = k, \qquad y_0 = y_{2m+1} = 0,$$

show that

$$y_n = \frac{1}{6}k\left[1 - (-1)^n \frac{\sin \mathrm{h}(2m + 1 - 2n)\,u}{\sin \mathrm{h}(2m + 1)\,u}\right] \quad \text{where} \quad u = \tfrac{1}{2}\ln\left(2 + \sqrt{3}\right).$$

5. Prove that the iterative formula

$$x_{n+1} = \frac{1}{2}\left(x_n + \frac{a}{x_n}\right)$$

converges to the square root of a for an arbitrary non-zero initial x_0; and that if x_0 is positive the limit is the positive square root and if x_0 is negative the limit is the negative square root.

Similarly show that the iterative formula

$$x_{n+1} = (1 \cdot 5 - 0 \cdot 5 \, ax_n^2) \, x_n$$

converges to $1/\sqrt{a}$ so that ax_n converges to the square root of a. The second formula is much more useful than the first in computers where division takes much longer than multiplication.

6. With a suitable initial choice x_0 prove that the iterative formula

$$x_{n+1} = x_n (2 - ax_n)$$

converges to $1/a$. In computers where there is no direct facility for division, this formula can be used for finding the reciprocal of a number a.

7. Find the quadratic factors of

$$x^4 + 13x^3 + 90x^2 + 300x + 642,$$

there being a factor near $x^2 + 4x + 17$. (U.L.)

[*Ans.* $(x^2 + 3 \cdot 809988x + 16 \cdot 822111)(x^2 + 9 \cdot 190012x + 38 \cdot 1640\,57)$]

8. Find the roots of

$$x^5 + x - 1 = 0. \quad \text{(U.L.)}$$

[*Ans.* $0 \cdot 754878, \, -0 \cdot 877439 \pm 0 \cdot 744862\,i, \, 0 \cdot 5 \pm 0 \cdot 866025\,i$]

9. Find the pair of complex roots with the largest modulus of the equation

$$x^4 + x^2 - x + 1 = 0. \quad \text{(U.L.)}$$

[*Ans.* $-0 \cdot 547424 \pm 1 \cdot 120873\,i$]

CHAPTER 5

Numerical Solution of Ordinary Differential Equations

THE NUMERICAL solution of a differential equation, e.g.

$$\frac{\mathrm{d}y}{\mathrm{d}x} = x + y^2, \qquad y = 0 \text{ when } x = 0,$$

means the computation of the values of y for various values of x, usually at equal intervals, e.g. for $x = 0 \cdot 1$ $(0 \cdot 1)$ 1. A mathematical solution usually means finding an explicit formula for y in terms of a finite number of elementary functions of x, for example, polynomial, trigonometric or exponential functions. A numerical solution is frequently possible when an explicit mathematical solution is not. Numerical methods usually apply to a larger variety of differential equations; thus the same methods can be used equally easily for linear and non-linear differential equations, whereas the latter are usually much more difficult to solve analytically than the former.

In the differential equation stated above, y is the *dependent* variable and x is the *independent* variable. It is of the first order because the first derivative is the highest derivative involved. The differential equation is said to be *ordinary* if there is one independent variable, and *partial* if there are two or more independent variables; this chapter deals with ordinary equations only and in Chapter 7 we discuss briefly one type of partial differential equation. If there is more than one dependent variable we have a *system* of differential equations; these are discussed in §5.3. We deal in §5.4 with some methods suitable for second order equations, but most of the rest of this chapter deals with the first order differential equation

$$\frac{\mathrm{d}y}{\mathrm{d}x} = f(x, y).$$

This is an important general type, and many other types can be reduced to this form.

Problems in differential equations can be classified according to the nature of the boundary conditions. If these conditions are all given at one point the problem is called an *initial value* problem or *marching* problem because the numerical solution is advanced step by step. If the boundary conditions are given at two or more points the problem is called a *boundary value* problem or *jury* problem. This chapter deals with initial value problems only; in Chapter 7 we deal briefly with some boundary value problems.

In this chapter we give only a small selection of the many numerical methods which exist, chosen to illustrate some of the main basic ideas. Discussion of errors, convergence, and various related problems is deferred to §§5.5–6.

5.1 Predictor–Corrector Methods

The problem is to solve the differential equation

$$\frac{\mathrm{d}y}{\mathrm{d}x} = f(x, y), \qquad y = y_0 \text{ when } x = x_0 \qquad (5.1.1)$$

at intervals h in x; for example, to solve the differential equation

$$\frac{\mathrm{d}y}{\mathrm{d}x} = x - y, \qquad y = 0 \text{ when } x = 0, \text{ at intervals } h = 0\cdot1.$$

One drawback of the methods of this section is that they require a few starting values of y: they are not "self starters" and we need some other method to find the first few values of y. For example, if $f(x, y)$ is easy to differentiate, the starting values can be obtained from Taylor's series:

$$y(x) = y(x_0) + (x - x_0) y'(x_0) + \tfrac{1}{2}(x - x_0)^2 y''(x_0) + \cdots$$

for $x = x_1 = x_0 + h$, $x_2 = x_0 + 2h$, $\ldots$

Thus in the above example we have

$$x_0 = 0 \qquad\qquad y_0 = 0$$
$$y'(x) = x - y \qquad y'(0) = 0$$
$$y''(x) = 1 - y' \qquad y''(0) = 1$$
$$y'''(x) = -y'' \qquad y'''(0) = -1$$
$$\cdots\cdots\cdots\cdots\cdots\cdots\cdots\cdots$$
$$y(x) = \tfrac{1}{2}x^2 - \tfrac{1}{6}x^3 + \ldots$$

We use this series to evaluate y for $x = 0\cdot1, 0\cdot2, 0\cdot3$ and use the given differential equation to evaluate y' at these points. Usually the series derived thus cannot be evaluated for more than a few steps beyond the starting point x_0 as then the series converges slowly or not at all.

The first method we explain is the Adams–Bashforth method. In this method we construct a table of

$$x, \ y, \ y' = f(x, y), \ \text{and differences of } '.$$

Suppose that we know y for $x = x_0, \ x_{-1}, \ x_{-2}, \ \ldots$. The problem is to advance the solution to $x = x_1$. We have

$$y_1 = y_0 + \int_{x_0}^{x_1} f(x, y) \, dx$$

We have to evaluate the integral in terms of $f_0 = f(x_0, y_0)$, $f_{-1} = f(x_{-1}, y_{-1}), \ \ldots$. From (3.9.2) we have

$$y_1 = y_0 + h(f_0 + \tfrac{1}{2}\nabla f_0 + \tfrac{5}{12}\nabla^2 f_0 + \tfrac{3}{8}\nabla^3 f_0 + \ldots). \qquad (5.1.2)$$

From (3.9.3) we have

$$y_1 = y_0 + h(f_1 - \tfrac{1}{2}\nabla f_1 - \tfrac{1}{12}\nabla^2 f_1 - \tfrac{1}{24}\nabla^3 f_1 - \ldots). \qquad (5.1.3)$$

The second formula is better because the coefficients are smaller, but we cannot use it immediately as f_1 and its backward differences are not known. The procedure is:

(1) Use (5.1.2) as a *predictor* to compute a predicted value $y_1^{(p)}$.

(2) Evaluate $f_1 = f(x_1, y_1^{(p)})$ and ∇f_1, $\nabla^2 f_1$, $\dots$.

(3) Use (5.1.3) to compute a corrected value $y_1^{(c)}$. Hence (5.1.3) is called a corrector.

(4) Evaluate f_1 and its backward differences again, but this time $f_1 = f(x_1, y_1^{(c)})$.

In theory we may have to repeat (3) and (4) several times until the new value of f_1 agrees with its previous value; in practice h is taken sufficiently small so as to make it unnecessary to apply the corrector more than once. In fact if we take a sufficient number of terms in (5.1.2) then usually $y_1^{(p)}$ and $y_1^{(c)}$ will agree and we can omit (3) and (4) and there is no need to evaluate f_1 more than once per step; it would be wise however to use (3) as a check.

Example. To solve the differential equation

$$y' = x - y, \quad y = 0 \text{ when } x = 0, \text{ for } x = 0\cdot1\,(0\cdot1)\,1.$$

By the method explained above we derive the series

$$y = \tfrac{1}{2}x^2 - \tfrac{1}{6}x^3 + \tfrac{1}{24}x^4 - \tfrac{1}{120}x^5$$

and evaluate this series for $x = \pm\,0\cdot1$, $\pm\,0\cdot2$, then use the differential equation to evaluate $y' = f(x, y) = x - y$. We obtain the part above the line in the following table:

x	y	f	∇f	$\nabla^2 f$	$\nabla^3 f$	$\nabla^4 f$
$-0\cdot2$	$0\cdot021403$	$-0\cdot221403$				
			116232			
$-0\cdot1$	$0\cdot005171$	$-0\cdot105171$		-11061		
			105171		1053	
$0\cdot0$	0	0		-10008		-102
			95163		951	
$0\cdot1$	$0\cdot004837$	$0\cdot095163$		-9057		-87
			86106		864	
$0\cdot2$	$0\cdot018731$	$0\cdot181269$		-8193		
			77913			
$0\cdot3$	$0\cdot040818$	$0\cdot259182$				

To advance the solution to $x = 0 \cdot 3$ we apply (5.1.2) taking $x_0 = 0 \cdot 2$

$$
\begin{aligned}
y_1 = y_0 \quad &= 0 \cdot 018731 \\
+ hf_0 \quad &= 0 \cdot 0181269 \\
+ \tfrac{1}{2}h\nabla f_0 \quad &= 0 \cdot 05 \times 0 \cdot 086106 \\
+ \tfrac{5}{12}h\nabla^2 f_0 \quad &= 0 \cdot 416667 \times -0 \cdot 0009057 \\
+ \tfrac{3}{8}h\nabla^3 f_0 \quad &= 0 \cdot 375 \times 0 \cdot 0000951 \\
+ \tfrac{251}{720}h\nabla^4 f_0 \quad &= 0 \cdot 348611 \times -0 \cdot 0000102 \\
= 0 \cdot 040818.
\end{aligned}
$$

Hence $f_1 = x - y_1 = 0 \cdot 259182$, and by differencing we obtain the remaining entries below the line in the table above. Substituting (5.1.3) we get the same value of y_1, thus there is no need for correction. We can now advance the solution to $x = 0 \cdot 4$ by taking $0 \cdot 3$ as x_0, and so on.

The second method we explain is the Milne–Simpson method. This uses (3.9.8) as a predictor and (3.9.7) as a corrector in the form

$$y_1 = y_{-3} + \tfrac{4}{3}h(2f_{-2} - f_{-1} + 2f_0), \qquad (5.1.4)$$

$$y_1 = y_{-1} + \tfrac{1}{3}h(f_{-1} + 4f_0 + f_1). \qquad (5.1.5)$$

Example. To solve the differential equation of the previous example, given the following table:

x	y	f
$-0 \cdot 1$	$0 \cdot 005171$	$-0 \cdot 105171$
0	0	0
$0 \cdot 1$	$0 \cdot 004837$	$0 \cdot 095163$
$0 \cdot 2$	$0 \cdot 018731$	$0 \cdot 181269$

We take $x_0 = 0 \cdot 2$, then (5.1.4) gives

$$
\begin{aligned}
y_1 &= 0 \cdot 005171 + \frac{0 \cdot 4}{3}(0 - 0 \cdot 095163 + 0 \cdot 362538) \\
&= 0 \cdot 040821,
\end{aligned}
$$

hence $\qquad f_1 = x_1 - y_1 = 0 \cdot 259179.$

Now we can apply (5.1.5) which gives

$$y_1 = 0 \cdot 004837 + \frac{0 \cdot 1}{3}\,(0 \cdot 095163 + 0 \cdot 72507 + 0 \cdot 259179)$$

$$= 0 \cdot 040818,$$

hence $\qquad f_1 = x_1 - y_1 = 0 \cdot 259182.$

The Milne–Simpson predictor corrector method is typical of several others of the same type introduced by W. E. Milne. Usually the corrector has to be applied at least once so that $f(x, y)$ has to be evaluated at least twice per step.

There are important differences between the two methods explained in this section. The Milne–Simpson method replaces the given differential equation

$$\frac{dy}{dx} = f(x, y)$$

by the finite difference approximation:

$$y_1 - y_0 = \tfrac{1}{3}h(f_{-1} + 4f_0 + f_1),$$

which is different from the given differential equation because we ignore a truncation term. We call this the truncated type; the finite difference equation which replaces the given equation has a solution which should closely resemble the solution of the given equation. The Adams–Bashforth method solves the given equation provided we take a sufficient number of terms in the predictor (5.1.2) and provided the round-off errors are negligible; we call this the untruncated type. The Adams–Bashforth method used in this way is not strictly a predictor corrector method as (5.1.3) is used only as a check. Both methods of this section are called multi-step methods because they require two or more preceding values of y. This is a disadvantage in cases when we may want to vary the step length h; we should take a small h when the solution varies rapidly and a large h when the solution varies slowly. Further, multi-step methods have a disadvantage because they are not self starters. The methods of the next section are single-step methods and do not have these disadvantages.

5.2 Single-Step Methods

The simplest of these is Euler's method:

$$y_1 = y_0 + hf(x_0, y_0)$$

where $\dfrac{dy}{dx} = f(x, y), \qquad y = y_0$ when $x = x_0$.

The truncation error in the formula is of order h^2 and h will have to be very small to get reasonable accuracy. This method is of the truncated type.

Next we explain the method of Taylor's series. It is of the single-step untruncated type. It is very powerful if there is some simple expression for the higher derivatives of y, usually in terms of the previous derivatives of y in the form of a recurrence relation:

$$y^{(n+1)}(x) = A(x)\, y^{(n)}(x) + B(x)\, y^{(n-1)}(x) + C(x).$$

Then at each step the solution is advanced by the formula

$$y_1 = y_0 + hy_0' + h^2 y_0''/2! + \ldots + h^n y_0^{(n)}/n! + \ldots,$$

taking the preceding point as (x_0, y_0) and as many terms as is necessary to attain full length accuracy. The higher derivatives are evaluated by the recurrence relation.

Example. To advance the solution of the differential equation

$$y' = x - y, \ y = 0 \cdot 018731 \text{ when } x = 0 \cdot 2, \ h = 0 \cdot 1.$$

We have $y'' = 1 - y'$ and $y^{(n+1)} = - y^{(n)}$,
hence

$$y'(0 \cdot 2) = \quad x - y \quad = \quad 0 \cdot 181269$$
$$y''(0 \cdot 2) = \quad 1 - y' \quad = \quad 0 \cdot 818731$$
$$y'''(0 \cdot 2) = - y''(0 \cdot 2) = - 0 \cdot 818731$$

. . . .

hence

$$
\begin{aligned}
y(0 \cdot 3) = y(0 \cdot 2) &&=&& 0 \cdot 018731 \\
+\ 0 \cdot 1 y'(0 \cdot 2) &&=&& 0 \cdot 0181269 \\
+\ 0 \cdot 01 y''(0 \cdot 2)/2! &&=&& 0 \cdot 0040937 \\
+\ 0 \cdot 001 y'''(0 \cdot 2)/3! &&=&& -\ 0 \cdot 0001365 \\
+\ 0 \cdot 0001 y^{iv}(0 \cdot 2)/4! &&=&& 0 \cdot 0000034 \\
= 0 \cdot 040818. &&&&
\end{aligned}
$$

The next method to be described, the Runge–Kutta method, is a truncated single-step method; but the truncation error is of order h^5. To advance the solution of the differential equation

$$
\frac{dy}{dx} = f(x, y), \qquad y = y_0 \text{ when } x = x_0
$$

to $x = x_1 = x_0 + h$, we compute successively:

$$
\begin{aligned}
k_1 &= hf(x_0, y_0), \\
k_2 &= hf(x_0 + \tfrac{1}{2}h, y_0 + \tfrac{1}{2}k_1), \\
k_3 &= hf(x_0 + \tfrac{1}{2}h, y_0 + \tfrac{1}{2}k_2), \\
k_4 &= hf(x_0 + h, y_0 + k_3).
\end{aligned}
$$

These four quantities are various estimates to

$$
y_1 - y_0 = \int_{x_0}^{x_1} f(x, y)\, dx.
$$

It can be shown that the weighted mean,

$$
\tfrac{1}{6}(k_1 + 2k_2 + 2k_3 + k_4),
$$

of these four quantities is a better estimate than any of them and has an error of order h^5. Hence

$$
y_1 = y_0 + \tfrac{1}{6}(k_1 + 2k_2 + 2k_3 + k_4). \qquad (5.2.1)
$$

5

Example. To advance the solution of the differential equation

$$\frac{dy}{dx} = x - y, \qquad y = 0\cdot018731 \text{ when } x = 0\cdot2,$$

to $x = 0\cdot3$. We have

$$k_1 = 0\cdot1(0\cdot2 - 0\cdot018731) \qquad = 0\cdot0181269$$
$$k_2 = 0\cdot1(0\cdot25 - 0\cdot027794) \qquad = 0\cdot0222206$$
$$k_3 = 0\cdot1(0\cdot25 - 0\cdot0298413) \qquad = 0\cdot0220159$$
$$k_4 = 0\cdot1(0\cdot3 - 0\cdot0407469) \qquad = 0\cdot0259253$$
$$y_1 = 0\cdot018731 + 0\cdot1325251/6 = 0\cdot040818$$

The Runge–Kutta method is popular on digital computers. It has the advantages of single-step methods: it does not require preceding values, and is therefore a self starter; it is easy to change the step length; and although it is of the truncated type, its truncation error is of high order. A possible disadvantage is the fact that $f(x, y)$ is evaluated four times for each step; this can be serious if $f(x, y)$ is complicated and lengthy to evaluate, whereas the predictor–corrector methods of the previous section do not require the evaluation of $f(x, y)$ more than once or twice per step. In such a case a suitable procedure would be to use the Runge–Kutta method to start the solution and to use the Adams–Bashforth or Milne–Simpson methods to continue the solution.

5.3 Systems of Differential Equations. Differential Equations of Order Higher than 1

So far we have been dealing with differential equations of the first order and of the form

$$\frac{dy}{dx} = f(x, y),$$

where x is the independent variable and y is the dependent variable. A system of differential equations of the first order has one independent variable x but several dependent variables

$(y_1, y_2, \ldots, y_n)$. It is of the form:

$$\frac{dy_1}{dx} = f_1(x, y_1, y_2, \ldots, y_n),$$

$$\frac{dy_2}{dx} = f_2(x, y_1, y_2, \ldots, y_n),$$

$$\cdots\cdots\cdots\cdots\cdots\cdots\cdots\cdots\cdots$$

$$\frac{dy_n}{dx} = f_n(x, y_1, y_2, \ldots, y_n).$$

To solve this system numerically we have to find the values of $(y_1, y_2, \ldots, y_n)$ at equal intervals $x_1 = x_0 + h$, $x_2 = x_0 + 2h$, $\ldots$ given the initial values of $(y_1, y_2, \ldots, y_n)$ at x_0. The methods of the previous two sections can be applied to several dependent variables in exactly the same way as to one dependent variable. The variables $(y_1, y_2, \ldots, y_n)$ are advanced simultaneously by steps of length h in x. The normal procedure on digital computers is to program the methods of the previous two sections for n dependent variables where n is a parameter which has to be specified; we also have to specify an *auxiliary sequence* which evaluates the functions $f_1, f_2, \ldots, f_n$, for a given $(x, y_1, y_2, \ldots, y_n)$. For the equation

$$\frac{dy}{dx} = f(x, y)$$

we specify $n = 1$.

To solve a differential equation of order n higher than 1, it is possible to reduce it to a system of differential equations of the first order in n dependent variables $(y_1, y_2, \ldots, y_n)$ defined as follows:

$$y_1 = y,$$

$$y_2 = \frac{dy}{dx},$$

$$y_3 = \frac{d^2y}{dx^2},$$

$$\cdots\cdots\cdots\cdots$$

$$y_n = \frac{d^{n-1}y}{dx^{n-1}}.$$

For example, the linear differential equation of order 3:

$$A \frac{d^3y}{dx^3} + B \frac{d^2y}{dx^2} + C \frac{dy}{dx} + Dy = E$$

where A, B, C, D, E may be functions of x, can be reduced to the following system:

$$y_1 = y,$$

$$\frac{dy_1}{dx} = f_1(x,\ y_1,\ y_2,\ y_3) = y_2,$$

$$\frac{dy_2}{dx} = f_2(x,\ y_1,\ y_2,\ y_3) = y_3,$$

$$\frac{dy_3}{dx} = f_3(x,\ y_1,\ y_2,\ y_3) = \frac{1}{A}(E - By_3 - Cy_2 - Dy_1).$$

The Runge–Kutta or predictor–corrector methods can then be applied. But for this special form of simultaneous differential equations it is possible to simplify the work when using the predictor–corrector methods. The predictor is applied to the last equation only, the remaining ones are integrated by the corrector formula.

Example. Given the following starting values for the differential equation

$$xy'' + y' + xy = 0,$$

x	y	y'	y''
0·0	1·0000	0·0000	−0·5000
0·1	0·9975	−0·0499	−0·4981
0·2	0·9900	−0·0995	−0·4925
0·3	0·9776	−0·1483	−0·4831

to advance the solution to $x = 0\cdot4$ we have, using the Milne–Simpson predictor to integrate y''

$$y'(0\cdot4) = y'(0) + \tfrac{4}{3}h[2y''(0\cdot1) - y''(0\cdot2) + 2y''(0\cdot3)]$$

$$= -0\cdot1960.$$

Using now the corrector to integrate y', we have

$$y(0\cdot4) = y(0\cdot2) + \tfrac{1}{3}h[y'(0\cdot2) + 4y'(0\cdot3) + y'(0\cdot4)] = 0\cdot9604.$$

Substituting in the differential equation we have

$$y'' = -\frac{1}{x}y' - y,$$

hence $\qquad\qquad\quad y''(0\cdot4) = -0\cdot4704.$

Using the corrector now to integrate y'' we get

$$y'(0\cdot4) = y'(0\cdot2) + \tfrac{1}{3}h[y''(0\cdot2) + 4y''(0\cdot3) + y''(0\cdot4)]$$
$$= -0\cdot1960$$

which is the same as before; otherwise the cycle of operations is repeated until there is agreement to 4 decimal places. Thus the solution is advanced by one step and we have the new line in the table:

$$0\cdot4 \qquad 0\cdot9604 \qquad -0\cdot1960 \qquad -0\cdot4704.$$

This method of solving differential equations of higher order by reducing them to a system of differential equations of the first order is popular on digital computers because one can use existing library programs for solving such systems. But there are more powerful methods suitable for solving equations of the second or higher order. We discuss two of these methods in the next section.

5.4 Further Methods

The methods explained in the previous three sections are general; they apply to all ordinary differential equations with initial boundary conditions. However, they are not always the best methods for particular forms of ordinary differential equations. In this section we deal with the following two forms:

$$y'' = f(x, y) \qquad\qquad (a)$$

i.e. a second order equation with the first derivative absent.

$$y'' + u(x)\, y' + v(x)\, y = w(x) \tag{b}$$

i.e. a linear second order equation where the coefficients may be functions of x.

A large number of differential equations which arise in the applied sciences have these forms or can be transformed to these forms.

The method for solving the first differential equation is called the Gauss–Jackson method. We use (3.8.4) in the form

$$\delta^2 y_n = h^2 y_n'' + \tfrac{1}{12} h^2 \delta^2 y_n'', \tag{5.4.1}$$

in which the error term is

$$- \tfrac{1}{240} h^2 \delta^4 y_n''$$

which is very small, of order h^6.

We tabulate y, $\delta^2 y$, y'', $\delta^2 y''$, the last items being

$$y_n, \qquad \delta^2 y_{n-1}, \qquad y_n'', \qquad \delta^2 y_{n-1}''.$$

To advance the solution to x_{n+1} we carry out the following steps:

(1) Guess $\delta^2 y_n''$ by any reasonable method. For example, we could assume that $\delta^2 y''$ is constant so that we take $\delta^2 y_n'' = \delta^2 y_{n-1}''$. A better guess would be given by assuming that $\delta^3 y''$ is constant so that we take

$$\delta^2 y_n'' = 2\delta^2 y_{n-1}'' - \delta^2 y_{n-2}''.$$

(2) Use (5.4.1) to calculate $\delta^2 y_n$. Hence calculate

$$y_{n+1} = 2y_n - y_{n-1} + \delta^2 y_n.$$

(3) Use the differential equation to calculate

$$y_{n+1}'' = f(x_{n+1}, y_{n+1}).$$

(4) Calculate $\delta^2 y_n''$ again this time using the formula

$$\delta^2 y_n'' = y_{n+1}'' - 2y_n'' + y_{n-1}'',$$

and hence compute e, the numerical value of the difference between the $\delta^2 y_n''$ computed here and the previous $\delta^2 y_n''$.

(5) Repeat the calculation from step (2) if $\frac{1}{12}h^2e$ contributes to the least significant figure.

For example, if $h = 0\cdot1$ and the work is carried to 6 decimal places then we have to repeat from step (2) if $\frac{1}{12}h^2e$ is greater than $0\cdot0000005$, i.e. if e is greater than $0\cdot000600$. This shows that the guess in step (1) can be fairly rough without having to repeat the calculation.

Example. To advance the solution of the differential equation

$$y'' = 1 + y - x$$

to $x = 0\cdot4$, given the following table:

x	y	$\delta^2 y$	y''	$\delta^2 y''$
$0\cdot0$	$0\cdot000000$		$1\cdot000000$	
$0\cdot1$	$0\cdot004837$	$0\cdot009057$	$0\cdot904837$	$0\cdot009057$
$0\cdot2$	$0\cdot018731$	$0\cdot008193$	$0\cdot818731$	$0\cdot008193$
$0\cdot3$	$0\cdot040818$		$0\cdot740818$	

(1) We take $\delta^2 y''(0\cdot3) = 2\delta^2 y''(0\cdot2) - \delta^2 y''(0\cdot1) = 0\cdot007329$.
(2) From (5.4.1) we get $\delta^2 y(0\cdot3) = 0\cdot007414$. Hence

$$y(0\cdot4) = 0\cdot070313.$$

(3) $y''(0\cdot4) = 1 + y - x = 0\cdot670319$.
(4) Hence $\delta^2 y''(0\cdot3) = 0\cdot007414$. The correction $e = 0\cdot000085$.
(5) There is no need to repeat the calculation.

The last two lines of the table will now be:

$0\cdot3$	$0\cdot040818$	$0\cdot007414$	$0\cdot740818$	$0\cdot007414$
$0\cdot4$	$0\cdot070319$		$0\cdot670319$	

This method is of the multi-step truncated type, the truncation error is of order h^6 and thus the method is more efficient than the Runge–Kutta or Milne–Simpson methods. One can make it even more powerful by taking more terms in (5.4.1), so that the method becomes of the untruncated type, calculating these extra terms as correction terms as in the method to be explained next.

In this way one can choose a much larger h without loss of accuracy. It may be possible to apply this method to differential equations of the second order involving the first derivative y' by applying first a suitable transformation of the dependent or independent variable which eliminates the first derivative.

The second method to be explained is the Fox–Goodwin method. This introduces the exceedingly important correction-term C which is a truncation error not ignored but used in a further approximation. This means that it is of the untruncated type; the final solution satisfies the original and not the truncated equation. The concept applies also to boundary problems, partial differential equations, and various other situations. We consider the differential equation

$$y'' + uy' + vy = w.$$

From (3.8.2) and (3.8.3) we have

$$y' = \frac{1}{h} (\mu\delta y - \tfrac{1}{6}\mu\delta^3 y + \ldots),$$

$$y'' = \frac{1}{h^2} (\delta^2 y - \tfrac{1}{12}\delta^4 y + \ldots).$$

Substituting these values in the differential equation we have

$$\delta^2 y + hu\mu\delta y + h^2 vy = h^2 w + C,$$

where C is a correction-term given by the expression

$$C = \tfrac{1}{12}\delta^4 y + \tfrac{1}{6}hu\mu\delta^3 y - \tfrac{1}{90}\delta^6 y - \tfrac{1}{30}hu\mu\delta^5 y + \ldots$$

We apply this equation at x_n, using

$$\delta^2 y_n = y_{n+1} - 2y_n + y_{n-1},$$

$$\mu\delta y_n = \tfrac{1}{2}(y_{n+1} - y_{n-1}),$$

and on simplifying we get

$$(1 + \tfrac{1}{2}hu) y_{n+1} - (2 - h^2 v) y_n + (1 - \tfrac{1}{2}hu) y_{n-1} = h^2 w + C,$$

or $$y_{n+1} = py_n + qy_{n-1} + r + s,$$

where
$$p = (2 - h^2 v)/(1 + \tfrac{1}{2}hu)$$
$$q = -(1 - \tfrac{1}{2}hu)/(1 + \tfrac{1}{2}hu)$$
$$r = h^2 w/(1 + \tfrac{1}{2}hu)$$
$$s = C/(1 + \tfrac{1}{2}hu).$$

At first s is ignored because it involves y_{n+1}, y_{n+2}, ... We thus solve the recurrence relation

$$y_{n+1} = py_n + qy_{n-1} + r \qquad (5.4.2)$$

for $n = 2, 3, 4, 5, \ldots$, say up to n = 12, given y_0 and y_1. Then we solve the recurrence relation

$$y_{n+1} = py_n + qy_{n-1} + r + s \qquad (5.4.3)$$

for $n = 2, 3, 4, \ldots$, 10, using y_0, y_1, and the y's obtained from (5.4.2) to calculate the correction-term s. We may have to apply (5.4.3) 2 or 3 times to ensure convergence.

Example. To continue the solution of

$$xy'' + y' + xy = 0$$

given the following starting values:

x	0	0·1	0·2	0·3	0·4	0·5
y	1·000000	0·997502	0·990025	0·977626	0·960398	0·938470

The equation can be put in the form

$$y'' + \frac{1}{x}y' + y = 0,$$

so that $$u = 1/x, \quad v = 1, \quad w = r = 0.$$

We tabulate below the results of applying (5.4.2) and (5.4.3).

x	p	q	y from (5.4.2)	s	y from (5.4.3)
0·6	1·836923	−0·846154	0·911996	0·000009	0·912006
0·7	1·857333	−0·866667	0·881178	0·000008	0·881203
0·8	1·872941	−0·882353	0·846243	0·000008	0·846290
0·9	1·885263	−0·894737	0·807453	0·000008	0·807526
1·0	1·895238	−0·904762	0·765096	0·000007	0·765199
1·1	1·903478	−0·913043	0·719486		

The error in the last column does not exceed 0.000003.

5.5 Accuracy: Error Build-up

The subject of error build-up is of prime importance but much of it is too advanced for this book. There are some general definitions and remarks at the end of this section and the next. But in order to appreciate the point of these definitions we consider as an illustration and in some detail one or two specific cases.

First we consider the Milne–Simpson method applied to the differential equation

$$\frac{\mathrm{d}y}{\mathrm{d}x} = f(x, y).$$

Let $z(x)$ be the true solution and $y(x)$ the result of the numerical solution defined at tabular points only. Let

$$z_n = z(x_n),$$

$$y_n = y(x_n).$$

The predictor formula (5.1.4) gives

$$z_{n+1} = z_{n-3} + \tfrac{4}{3}h(2f_{n-2} - f_{n-1} + 2f_n) + T_1,$$

where

$$T_1 = \tfrac{28}{90}h^5 f^{\mathrm{iv}}(t_1), \text{ for } x_{n-3} \leqq t_1 \leqq x_{n+1}.$$

The predicted y denoted by $y^{(p)}$ or $y_{n+1}^{(p)}$ is given by

$$y_{n+1}^{(p)} = y_{n-3} + \tfrac{4}{3}h(2f_{n-2} - f_{n-1} + f_n).$$

The corrector formula (5.1.5) gives

$$z_{n+1} = z_{n-1} + \tfrac{1}{3}h(f_{n-1} + 4f_n + f_{n+1}) + T_2 \qquad (5.5.1)$$

where

$$T_2 = -\frac{h^5}{90} f^{iv}(t_2), \text{ for } x_{n-1} \leqq t_2 \leqq x_{n+1}.$$

The corrected y denoted by $y^{(c)}$ or $y_{n+1}^{(c)}$ is given by

$$y_{n+1}^{(c)} = y_{n-1} + \tfrac{1}{3}h(f_{n-1} + 4f_n + f_{n+1}). \qquad (5.5.2)$$

It follows that

$$T_1 \sim - 28T_2.$$

Assuming that there are no previous errors we have

$$y^p + T_1 \sim y^c + T_2 \sim z_{n+1},$$

hence

$$T_2 \sim \tfrac{1}{29}(y^p - y^c). \qquad (5.5.3)$$

This is a useful estimate of T_2 which is called the *error per step*. Let

$$z_n = y_n + e_n$$

where e_n is the error in the solution at the nth step. This is different from the error per step T_2; e_n is the result of the build-up of errors from the nth step and all the previous steps.

We now apply the above equations to the solution of a particular differential equation:

$$\frac{\mathrm{d}y}{\mathrm{d}x} = Ay + Bx + C$$

where A, B, C are constants. This special differential equation is, however, important because in the general differential equation, $f(x, y)$ can be approximated locally, possibly after a shift of origin, in the form

$$f(x, y) \sim Ay + Bx + C.$$

Because the Simpson rule integrates exactly the first order polynomial $Bx + C$ we can omit it when studying the error build-up due to truncation errors. Thus let $z_n = z(x_n)$ be the exact solution and y_n the numerical solution of the differential equation

$$\frac{dy}{dx} = Ay \qquad (5.5.4)$$

where
$$x_0 = 0, \qquad x_n = nh.$$
If we substitute
$$f_n = Az_n,$$
$$z_n = y_n + e_n,$$
in (5.5.1), and
$$f_n = Ay_n$$

in (5.5.2), and subtract the resulting two equations we get

$$y_{n+1} = y_{n-1} + \tfrac{1}{3}Ah(y_{n-1} + 4y_n + y_{n+1}), \qquad (5.5.5)$$

$$e_{n+1} = e_{n-1} + \tfrac{1}{3}Ah(e_{n-1} + 4e_n + e_{n+1}) + T_2. \qquad (5.5.6)$$

We observe that T_2 is of order h^5 and is negligible as compared with the other coefficients in (5.5.6). We are interested in the asymptotic behaviour of e_n when e_n may be large. We can therefore omit the term T_2 in order to solve the recurrence relation (5.5.6). We also observe that the numerical solution y_n satisfies essentially the same recurrence relation as the error e_n. This is usually the case in multi-step methods.

We write (5.5.6) in the following form:

$$(1 - \tfrac{1}{3}Ah)\, e_{n+1} - \tfrac{4}{3}Ahe_n - (1 + \tfrac{1}{3}Ah)\, e_{n-1} = 0.$$

We can solve this recurrence relation by the methods of §4.1. We get

$$e_n = pu^n + qv^n$$

where u, v are the roots of the quadratic equation

$$(1 - \tfrac{1}{3}Ah)\, x^2 - \tfrac{4}{3}Ahx - (1 + \tfrac{1}{3}Ah) = 0.$$

By binomial expansion in powers of h we get

$$u \sim 1 + Ah, \qquad v \sim - (1 - \tfrac{1}{3}Ah)$$
and
$$u^n \sim (1 + Ah)^n \sim e^{Anh} = e^{Ax_n};$$

similarly $\qquad\qquad v^n \sim (-1)^n e^{-\frac{1}{2}Ax_n},$

hence $\qquad\qquad e_n \sim pe^{Ax_n} + (-1)^n qe^{-Ax_n}.$ $\qquad$ (5.5.7)

On the other hand from the explicit solution of (5.5.4) we have

$$y_n \sim z_n = Pe^{Ax_n}, \qquad (5.5.8)$$

where P depends upon initial conditions. If $P \neq 0$, then p and q are usually small compared with P. We can now study the asymptotic effect of e_n. We distinguish two cases:

(a) A is positive. The solution (5.5.8) grows rapidly in magnitude. The second term in (5.5.7) is negligible compared with the first term. We have
$$e_n \sim pe^{Ax_n}$$

and although the error also grows rapidly, it remains a small proportion p/P of the solution.

(b) A is negative. The solution (5.5.8) decreases in magnitude as n increases. The first term in (5.5.7) is negligible. We have

$$e_n \sim (-1)^n qe^{-\frac{1}{2}Ax_n}$$

The error increases in magnitude and oscillates in sign. The error will swamp the solution and the procedure is said to be *unstable*. This kind of unstability which arises when the true solution is decreasing can be avoided sometimes by solving the differential equation by negative steps in x, but because of the initial condition this may not be easy.

There is an important way of looking at the second term in (5.5.7). The numerical procedure for solving the first order differential equation (5.5.4) replaces it by the second order recurrence relation (5.5.5). This recurrence relation has two independent solutions: Pu^n, Qv^n; the first behaves like the solution of the given equation, the other is spurious. If this spurious solution dies out the method is stable, otherwise it is unstable. This situation occurs whenever a differential equation is solved by a multi-step truncated method which reduces to a recurrence relation of an order higher than the order of the given differential equation.

Example. The following example illustrates the unstable case (b) above where A is negative and the error oscillates in sign and increases in relative magnitude. We list x_n, z_n, y_n, e_n for the differential equation

$$y' = x - y, \qquad y = 0 \text{ when } x = 0$$

The entry z_n is obtained from the known explicit solution

$$z = e^{-x} + x - 1$$

The entry y_n is obtained by the Milne–Simpson method. The arithmetic was carried out on a digital computer to 29 binary significant figures, i.e. to about $8 \cdot 5$ decimal significant figures. For comparison one table is given for $h = 0 \cdot 1$ and another for $h = 0 \cdot 5$. As the nature of the error e_n emerges slowly we give few entries for small n and then a few entries for large n. When n is large, it is clear that the errors alternate in sign and increase in value.

$$h = 0 \cdot 1$$

n	x	z	y	e
4	0·4	0·07032005	0·07031997	0·00000008
5	0·5	0·10653066	0·10653060	0·00000006
6	0·6	0·14881164	0·14881150	0·00000013
7	0·7	0·19658531	0·19658521	0·00000010
.	...	....	....	....
89	8·9	7·90013620	7·90013663	−0·00000043
90	9·0	8·00012323	8·00012269	0·00000054
91	9·1	8·10011151	8·10011199	−0·00000048
92	9·2	8·20010092	8·20010033	0·00000059

$$h = 0 \cdot 5$$

4	2·0	1·13533529	1·13526809	0·00006719
5	2·5	1·58208500	1·58208265	0·00000236
6	3·0	2·04978708	2·04971572	0·00007136
7	3·5	2·53019738	2·53022151	−0·00002412
.	...	....	....	....
27	13·5	12·50000139	12·50126495	−0·00126356
28	14·0	13·00000085	12·99851264	0·00148821
29	14·5	13·50000050	13·50175353	−0·00175303
30	15·0	14·00000030	13·99793561	0·00206470

We consider now the types of error which may occur when solving a differential equation.

(i) Truncation error arising from the use of approximate formulae. This type of error is absent in untruncated methods such as the Taylor series method or the Adams–Bashforth method.

(ii) Round-off errors arising from the limited number of significant figures. This error is random whereas the truncation error is systematic and usually larger.

(iii) Inherent errors due, for example, to approximate boundary values.

(iv) Blunders to be found and removed.

We consider next the way errors build up:

(a) Errors can die out so that only the last few matter.
(b) Errors such as rounding errors can build up statistically. This is not too serious as they grow like the square root of the number of steps.
(c) Errors can grow exponentially and this may or may not cause instability depending upon whether the solution is decreasing or increasing in magnitude.

We illustrate some of these ideas by considering the numerical solution of a second order differential equation:

$$y'' = ky \text{ where } k = m^2, \ m > 0.$$

The solution is of the form

$$y = Ae^{mx} + Be^{-mx}.$$

If the boundary conditions are such that $A = 0$ then we have *inherent instability*, a component of e_n will swamp the true solution. If $k = -m^2$ then the solution is trigonometric:

$$y = A \sin mx + B \cos mx.$$

Then errors accumulate statistically and remain small provided the numerical method does not introduce a spurious exponential increasing solution.

Finally we conclude with some definitions concerning the numerical solution of differential equations. A method is said to be *convergent* if the numerical solution tends to the true solution when h tends to zero. A method is said to be *stable* if as n tends to infinity y_n has the same behaviour as the true solution. Stability may depend upon the method; untruncated methods are more stable, some methods can be unstable regardless of the particular equation. Stability may depend upon the particular equation and initial values. Finally stability may depend upon taking h sufficiently small.

5.6 Solution of Differential Equations on Digital Computers

Differential equations of the first order are important as was explained in §5.4 and any particular method is often programmed for a system of n differential equations in n dependent variables where n is a parameter which has to be specified. The differential equation $\mathrm{d}y/\mathrm{d}x = f(x, y)$ is the case $n = 1$. The program requires an auxiliary sequence for the calculation of the expressions

$$\frac{\mathrm{d}y_i}{\mathrm{d}x} = f_i(x, y_1, y_2, \ldots, y_n)$$

given $(x, y_1, y_2, \ldots, y_n)$. This makes it possible for the same library program to be used for any system of differential equations; only the auxiliary sequence has to be changed. Given the step length h and the initial values of $(x, y_1, \ldots, y_n)$ the program replaces them by $x + h$ and the corresponding values of $y_1, y_2, \ldots, y_n$ at $x + h$. This procedure is particularly suitable for single-step methods. If there is an easy formula for the error per step such as (5.5.3) then it is possible to devise a more versatile variable step program. A certain desired accuracy E is specified; if the error per step is numerically greater than E then the step length h is halved, if it is much less than E then h is doubled. The effect of this simple scheme is that h expands when the solution is slowly-varying and contracts when the solution is approaching rapid oscillations or near a singularity where the solution varies rapidly. Again, this is suitable for single-step methods; it is much more difficult to vary the step length in multi-step methods.

Whether a formula for the error per step is available or not, it is not easy to detect the effect of error build-up over a large number of steps; this could be done by repeating the computation using double length arithmetic or slightly different initial conditions. For awkward differential equations a detailed error analysis along the lines of the previous section may be necessary and special methods may have to be used to take advantage of the properties of a particular equation. For example if, for a particular differential equation, there is an easy recurrence relation for computing higher derivatives, then the Taylor series method is very powerful. It is tempting to use an existing general library program; and it is easy on digital computers to get solutions with few or no accurate figures without being aware of it. One should always have some check on the credibility of the results. A danger signal is a rapidly varying solution or an appreciably different result when the solution is carried out for a slightly different interval.

EXERCISES

1. Show by successive approximations or otherwise that the solution of the differential equation

$$y' = 1 + xy + x^2y^2, \text{ where } y = 0 \text{ when } x = 0,$$

is

$$y = x + \tfrac{1}{3}x^3 + \tfrac{4}{15}x^5 + \tfrac{2}{15}x^7 + \ldots.$$

Use this series to calculate to 5 decimal places the values of y for $x = 0 \cdot 1 \ (0 \cdot 1) \ 0 \cdot 3$. Hence obtain by a step-by-step process, including checks, the value of y when $x = 0 \cdot 4$.   (U.L.)

[*Ans.* 0·100336 0·202754 0·309679 0·424307]

2. Solve the differential equation of Exercise 1 by the Taylor series method.

[*Hint:* Let $z = y^2$ and let $y^{(n)}$ denote the nth derivative of y. We have the following recurrence relations

$$y^{(n+1)} = xy^{(n)} + ny^{(n-1)} + x^2z^{(n)} + 2nxz^{(n-1)} + n(n-1)z^{(n-2)}$$

$$z^{(n)} = yy^{(n)} + \binom{n}{1} y'y^{(n-1)} + \binom{n}{2} y''y^{(n-2)} + \dots \ ]$$

3. Show that the differential equation

$$y'' + \sin y = 0 \qquad y = 0 \text{ and } y' = 1 \text{ when } x = 0$$

is satisfied by the series

$$y = x - x^3/3! + 2x^5/5! - 13x^7/7! + \dots.$$

Use this series to evaluate y to 6 decimal places for $x = 0 \ (0 \cdot 1) \ 0 \cdot 4$. Continue the solution by a step-by-step method for $x = 0 \cdot 5 \ (0 \cdot 1) \ 1 \cdot 8$. Prove that the maximum value of y is $\frac{1}{3}\pi$ and find from the numerical solution the corresponding value of x.

[*Hint:* The solution in the neighbourhood of the maximum is

x	$1 \cdot 5$	$1 \cdot 6$	$1 \cdot 7$	$1 \cdot 8$
y	$1 \cdot 032277$	$1 \cdot 044012$	$1 \cdot 047107$	$1 \cdot 041546$]

4. Given the differential equation

$$y' = 1 + 2xy \qquad y = 0 \text{ when } x = 0$$

show that the series expansion of y in powers of x as far as x^5 is

$$y = x + \tfrac{2}{3}x^3 + \tfrac{4}{15}x^5.$$

Hence tabulate y and y' to 4 decimal places for $x = 0 \ (0 \cdot 1) \ 0 \cdot 3$ and apply the Adams–Bashforth method to find y when $x = 0 \cdot 4$. (B.C.)

[*Ans.* $0 \cdot 4455$]

5. Establish a recurrence relation for higher derivatives in the equation of Exercise 4 and continue the solution by the Taylor series method.

6. Obtain the formula

$$y_2 = 2y_1 - y_0 + \tfrac{1}{12}h^2(f_0 + 10f_1 + f_2)$$

for the step-by-step integration of the differential equation

$$y'' = f(x, y).$$

Apply this method to solve the differential equation

$$y'' + xy = 0 \qquad y(0) = 0 \cdot 35503 \qquad y(0 \cdot 1) = 0 \cdot 38085$$

for $x = 0 \cdot 2$ and $0 \cdot 3$. (B.C.)

[*Ans.* $0 \cdot 40628, \ 0 \cdot 43089$]

7. Given the differential equation

$$y'' + xy = 0, \qquad y(0) = 1, \qquad y'(0) = 0,$$

show that the series expansion of y in powers of x as far as the term in x^6 is

$$y = 1 - \tfrac{1}{6}x^3 + \tfrac{1}{180}x^6.$$

Hence tabulate to 4 decimal places y, y' and y'' for $x = 0\ (0\cdot1)\ 0\cdot3$ and use the Adams–Bashforth method to find y and y' when $x = 0\cdot4$.   (B.C.)

CHAPTER 6

Matrices

6.1 Elementary Properties

This section is mostly a review of basic properties of matrices. We deal with three concepts: scalars, vectors, and matrices.

A *scalar* is a real or complex number.

A *vector* is a one-dimensional array of scalars

$$\mathbf{x} = (x_1, x_2, \ldots, x_n).$$

The scalar x_i is said to be the ith component of $\mathbf{x}$, and n, the number of components, is the dimension of the vector $\mathbf{x}$.

Given two vectors

$$\mathbf{x} = (x_1, x_2, \ldots, x_n),$$
$$\mathbf{y} = (y_1, y_2, \ldots, y_n)$$

we define the vector $p\mathbf{x} + q\mathbf{y}$, where p, q are scalars, by the equation

$$p\mathbf{x} + q\mathbf{y} = (px_1 + qy_1, \ px_2 + qy_2, \ \ldots, \ px_n + qy_n).$$

This defines the meaning of vectors such as

$$\mathbf{x} + \mathbf{y}, \ \mathbf{x} - \mathbf{y}, \ p\mathbf{x}, \ \ldots$$

We define the *scalar product* or *dot product* $\mathbf{x} \cdot \mathbf{y}$ of the vectors $\mathbf{x}$, $\mathbf{y}$ by the equation

$$\mathbf{x} \cdot \mathbf{y} = x_1 y_1 + x_2 y_2 + \ldots + x_n y_n.$$

We note that, for example, the vector expression $\mathbf{x} + \mathbf{y}$ and the scalar expression $\mathbf{x} \cdot \mathbf{y}$ are defined only if $\mathbf{x}$ and $\mathbf{y}$ have the same dimension.

If $\mathbf{x}$ and $\mathbf{y}$ have real components, they are said to be *orthogonal* if $\mathbf{x} \cdot \mathbf{y} = 0$, and $\mathbf{x}$ is said to be a *unit vector* if $\mathbf{x} \cdot \mathbf{x} = 1$.

A *matrix* is a two-dimensional array of scalars; this is equivalent to a one-dimensional array of vectors. A matrix **A** can be written in the form

$$\mathbf{A} = (a_{ij}) = \begin{pmatrix} a_{11} & a_{12} & \cdots & a_{1n} \\ a_{21} & a_{22} & \cdots & a_{2n} \\ \cdot & \cdot & \cdots & \cdot \\ a_{m1} & a_{m2} & \cdots & a_{mn} \end{pmatrix}.$$

Here a_{ij} denotes the element in the ith row and jth column. The matrix **A** is said to be an $(m \times n)$ matrix; it has m rows and n columns. It is made up of m row vectors each of dimension n, or it can be considered to be made up of n column vectors each of dimension m. The transpose of a matrix **A** is obtained by interchanging rows and columns, it is denoted by $\mathbf{A}^T$; for example

$$\text{if } \mathbf{A} = \begin{pmatrix} 3 & 1 & 5 \\ 2 & 4 & 6 \end{pmatrix} \text{ then } \mathbf{A}^T = \begin{pmatrix} 3 & 2 \\ 1 & 4 \\ 5 & 6 \end{pmatrix},$$

so that the ith row of **A** becomes the ith column of $\mathbf{A}^T$. In particular we shall often deal with a column vector

$$\mathbf{x} = \begin{pmatrix} x_1 \\ x_2 \\ \cdot \\ x_n \end{pmatrix}$$

which for typographical reasons is specified in its transpose form

$$\mathbf{x}^T = (x_1, x_2, \ldots, x_n)$$

or

$$\mathbf{x} = (x_1, x_2, \ldots, x_n)^T.$$

If **A**, **B** are two ($m \times n$) matrices, then we define the matrix

$$\mathbf{C} = (c_{ij}) = p\mathbf{A} + q\mathbf{B}$$

by the equations

$$c_{ij} = pa_{ij} + qb_{ij}.$$

This defines the meaning of matrix expressions such as

$$\mathbf{A} + \mathbf{B}, \ \mathbf{A} - \mathbf{B}, \ p\mathbf{A}, \ \ldots$$

If **A** is an ($m \times n$) matrix and **B** is an ($n \times 1$) matrix, we define the matrix product

$$\mathbf{C} = \mathbf{A} \times \mathbf{B} \ \text{ or } \ \mathbf{AB} = (c_{ij})$$

as follows: let

$$r_i(\mathbf{A}) \text{ denote the } i\text{th row of } \mathbf{A}$$

$$c_j(\mathbf{B}) \text{ denote the } j\text{th column of } \mathbf{B}$$

then

$$c_{ij} = r_i(\mathbf{A}) \ . \ c_j(\mathbf{B})$$

$$= \text{scalar product of the } i\text{th row of } \mathbf{A} \text{ and } j\text{th column of } \mathbf{B}.$$

Thus **C** is an ($m \times l$) matrix. We note that multiplication is only possible when the number of columns in **A** is the same as the number of rows in **B**, so that if the product **AB** is possible the product **BA** may not be possible and even if the second product is possible it is usually different from the first that is,

$$\mathbf{AB} \neq \mathbf{BA}$$

Example. Let

$$\mathbf{A} = \begin{pmatrix} 2 & 1 & -3 \\ 3 & 1 & 4 \end{pmatrix}, \qquad \mathbf{B} = \begin{pmatrix} 1 & -1 & 2 & 4 \\ 3 & 1 & 1 & 0 \\ 2 & -1 & 1 & 3 \end{pmatrix};$$

then

$$\mathbf{C} = \mathbf{A} \times \mathbf{B} = \begin{pmatrix} -1 & 2 & 2 & -1 \\ 14 & -6 & 11 & 24 \end{pmatrix}.$$

For instance,

$$c_{23} = r_2(\mathbf{A}) \cdot c_3(\mathbf{B}) = (3,\ 1,\ 4) \cdot (2,\ 1,\ 1)^{\mathrm{T}}$$
$$= 3 \times 2 + 1 \times 1 + 4 \times 1 = 11.$$

A matrix $\mathbf{A}$ is a *square matrix* if it has the same number of rows as of columns; it is symmetric if

$$\mathbf{A}^{\mathrm{T}} = \mathbf{A}, \text{ i.e. } a_{ij} = a_{ji}.$$

The *diagonal* of a matrix consists of the elements a_{ii}. An important square matrix is the *unit matrix* $\mathbf{I}$ which has units along the diagonal and zeros everywhere else; thus if the number of rows or columns is 3 then

$$\mathbf{I} = \begin{pmatrix} 1 & 0 & 0 \\ 0 & 1 & 0 \\ 0 & 0 & 1 \end{pmatrix}.$$

The unit matrix has many properties similar to those of the unit for scalars, for example,

$$\mathbf{IA} = \mathbf{AI} = \mathbf{A}.$$

The *inverse* matrix $\mathbf{A}^{-1}$ of a square matrix $\mathbf{A}$ is such that

$$\mathbf{A}^{-1}\mathbf{A} = \mathbf{A}\mathbf{A}^{-1} = \mathbf{I}.$$

For example, if we consider the two matrices

$$\mathbf{A} = \begin{pmatrix} 2 & 1 & 3 \\ 1 & -1 & 1 \\ 1 & -2 & 1 \end{pmatrix} \quad \mathbf{A}^{-1} = \begin{pmatrix} -1 & 7 & -4 \\ 0 & 1 & -1 \\ 1 & -5 & 3 \end{pmatrix}$$

we can verify that $\mathbf{AA}^{-1} = \mathbf{A}^{-1}\mathbf{A} = \mathbf{I}$. We will consider later the condition for the existence and the method of deriving the inverse.

A set of vectors $\mathbf{x}$, $\mathbf{y}$, $\mathbf{z}$, $\ldots$, is said to be *linearly dependent* if there is a linear relation between two or more of the vectors, i.e.

if one of the vectors is a linear combination of the others. If no such relation can be found then the set of vectors is said to be *linearly independent*. For example, if

$$\mathbf{x} = (2, \ 1, \ 5)$$
$$\mathbf{y} = (1, \ -1, \ 1)$$
$$\mathbf{z} = (0, \ 1, \ 1)$$

we can verify that

$$\mathbf{x} = 2\mathbf{y} + 3\mathbf{z}$$

thus $\mathbf{x}$ is linearly dependent on $\mathbf{y}$ and $\mathbf{z}$.

Consider an $(m \times n)$ matrix $\mathbf{A}$ as a set of m row vectors. If we can find from among these a set of r linearly independent vectors such that each of the remaining $m - r$ vectors can be expressed as some linear combination of the set of r vectors then we say that $\mathbf{A}$ has *rank r*. The rank of a matrix can be defined either in terms of its row vectors or in terms of its column vectors; both ways give the same rank.

Examples.

$$\mathbf{I} = \begin{pmatrix} 1 & 0 & 0 \\ 0 & 1 & 0 \\ 0 & 0 & 1 \end{pmatrix} \quad \text{has rank 3}$$

$$\mathbf{A} = \begin{pmatrix} 1 & -1 & 1 \\ 0 & 1 & 1 \\ 2 & 1 & 5 \end{pmatrix} \quad \text{has rank 2 because}$$
$$r_3(\mathbf{A}) = 2r_1(\mathbf{A}) + 3r_2(\mathbf{A})$$

$$\mathbf{A} = \begin{pmatrix} 1 & -1 & 1 \\ 2 & -2 & 2 \\ 3 & -3 & 3 \end{pmatrix} \quad \text{has rank 1 because}$$
$$r_2 = 2r_1 \text{ and } r_3 = 3r_1$$

To each square matrix **A** corresponds a scalar quantity:

$$\det \mathbf{A} = \text{determinant of } \mathbf{A}.$$

It is defined recursively in the following way:

For a (1×1) matrix $\mathbf{A} = (a_{11})$, $\det \mathbf{A} = a_{11}$.

For a (2×2) matrix $\mathbf{A} = \begin{pmatrix} a_{11} & a_{12} \\ a_{21} & a_{22} \end{pmatrix}$,

$$\det \mathbf{A} = a_{11}a_{22} - a_{21}a_{12}.$$

For a (3×3) matrix $\mathbf{A} = \begin{pmatrix} a_{11} & a_{12} & a_{13} \\ a_{21} & a_{22} & a_{23} \\ a_{31} & a_{32} & a_{33} \end{pmatrix}$,

we can expand $\det \mathbf{A}$ in terms of any row or column. Expanding in terms of the first row we define

$$\det \mathbf{A} = a_{11}A_{11} + a_{12}A_{12} + a_{13}A_{13}$$

where $A_{ij} = (-1)^{i+j}$ the determinant of the matrix derived from **A** by omitting the ith row and jth column.

A_{ij} is said to be the *cofactor* of a_{ij}. Thus

$$A_{11} = \det \begin{pmatrix} a_{22} & a_{23} \\ a_{32} & a_{33} \end{pmatrix}$$

$$A_{12} = -\det \begin{pmatrix} a_{21} & a_{23} \\ a_{31} & a_{33} \end{pmatrix} \qquad A_{13} = \det \begin{pmatrix} a_{21} & a_{22} \\ a_{31} & a_{32} \end{pmatrix}$$

For an $(n \times n)$ matrix $\mathbf{A} = (a_{ij})$, if we expand in terms of the ith row we have

$$\det \mathbf{A} = a_{i1}A_{i1} + a_{i2}A_{i2} + \ldots + a_{in}A_{in}$$

Thus it is expressed in terms of determinants of order $n - 1$ each of which is expressed in terms of determinants of order $n - 2$

and so on. We observe that this definition is numerically impractical and inaccurate because it leads to a sum of a large number of terms which largely cancel out. Fortunately, the evaluation of determinants is frequently unnecessary and we give below a more suitable method for evaluating the determinant of a general matrix.

We list some general properties of determinants expressed in terms of rows, but they hold equally well when expressed in terms of columns.

The square ($n \times n$) matrix $\mathbf{A}$ is said to be *singular* if det $\mathbf{A} = 0$. This happens if and only if the rows of $\mathbf{A}$ are linearly dependent, i.e. if rank $(\mathbf{A}) < n$.

If two rows of $\mathbf{A}$ are interchanged then det $\mathbf{A}$ changes sign.

det $\mathbf{A}$ is unaltered if we add to a row a linear combination of the other rows—these being left unchanged.

Finally, det $(\mathbf{AB}) = $ det $\mathbf{A} \times$ det $\mathbf{B}$.

A square matrix $\mathbf{U} = (u_{ij})$ is said to be *upper triangular* if $u_{ij} = 0$ when $i > j$. Thus for a (3×3) matrix we have

$$\mathbf{U} = \begin{pmatrix} u_{11} & u_{12} & u_{13} \\ 0 & u_{22} & u_{23} \\ 0 & 0 & u_{33} \end{pmatrix}.$$

For such a matrix

$$\det \mathbf{U} = u_{11}u_{22}u_{33} \ldots u_{nn}$$

The determinant is the product of the diagonal terms. This gives a practical way for evaluating det $\mathbf{A}$: by the method of elimination to be explained in the next section, a square matrix $\mathbf{A}$ is reduced to an upper triangular matrix $\mathbf{U}$ so that det $\mathbf{A} = $ det $\mathbf{U}$.

A set of n linear equations in n unknowns

$$\left. \begin{aligned} a_{11}x_1 + a_{12}x_2 + \ldots + a_{1n}x_n &= b_1, \\ a_{21}x_1 + a_{22}x_2 + \ldots + a_{2n}x_n &= b_2, \\ \ldots \quad\quad \ldots \quad\quad\quad\quad \ldots \\ a_{n1}x_1 + a_{n2}x_2 + \ldots + a_{nn}x_n &= b_n, \end{aligned} \right\} \quad (6.1.1)$$

can be put in the matrix form

$$\mathbf{Ax} = \mathbf{b}, \qquad (6.1.2)$$

where $\mathbf{A}$ is the matrix of the coefficients in (6.1.1), $\mathbf{x}$ is the column vector of the unknowns:

$$\mathbf{x} = (x_1, x_2, \ldots, x_n)^{\mathrm{T}}$$

and $\mathbf{b}$ is the column vector of the right-hand sides in (6.1.1):

$$\mathbf{b} = (b_1, b_2, \ldots, b_n)^{\mathrm{T}}.$$

$\mathbf{Ax}$ is the matrix product of $\mathbf{A}$ and $\mathbf{x}$, the vector $\mathbf{x}$ being considered as an $(n \times 1)$ matrix. An explicit formula for the solution of (6.1.1) is given by Cramer's rule:

$$\frac{x_1}{\det \mathbf{A}_1} = \frac{x_2}{\det \mathbf{A}_2} = \cdots = \frac{x_n}{\det \mathbf{A}_n} = \frac{1}{\det \mathbf{A}},$$

where $\mathbf{A}_i$ is the matrix derived from $\mathbf{A}$ by replacing its ith column by the column vector $\mathbf{b}$. This method of solving (6.1.1) is numerically impractical for $n > 3$.

If $\mathbf{A}^{-1}$, the inverse of $\mathbf{A}$, is known, then (6.1.1) can be solved explicitly in the form $\mathbf{x} = \mathbf{A}^{-1}\mathbf{b}$.

Example. Let

$$\mathbf{A} = \begin{pmatrix} 2 & 1 & 3 \\ 1 & -1 & 1 \\ 1 & -2 & 1 \end{pmatrix}, \qquad \mathbf{b} = \begin{pmatrix} 6 \\ 1 \\ 0 \end{pmatrix};$$

we can verify that

$$\mathbf{A}^{-1} = \begin{pmatrix} -1 & 7 & -4 \\ 0 & 1 & -1 \\ 1 & -5 & 3 \end{pmatrix}$$

and that the solution of $\mathbf{Ax} = \mathbf{b}$ is $\mathbf{x} = \mathbf{A}^{-1}\,\mathbf{b} = (1, 1, 1)^{\mathrm{T}}.$

But if the inverse matrix is not known then this is not a practical way of solving (6.1.1) as the computation of the inverse matrix is much longer than the direct solution of (6.1.1) by the method explained in the next section.

We give now an explicit formula for the inverse A^{-1} of a square matrix $A = (a_{ij})$. As before we define the cofactor of a_{ij} denoted by A_{ij}, as the product of $(-1)^{i+j}$ and the determinant of the $(n-1 \times n-1)$ matrix derived from A by omitting the row and column through a_{ij}. We define the *adjoint* matrix

$$A^* = (b_{ij}) \text{ where } b_{ij} = A_{ji},$$

so that the jth column of A^* consists of the cofactors of the elements of the jth row of A. Hence

$$r_i(A) \, . \, c_j(A^*) = a_{i1}A_{j1} + a_{i2}A_{j2} + \ldots + a_{in}A_{jn} = \det A',$$

where A' is derived from A by replacing its jth row by its ith row.

Hence
$$r_i(A) \, . \, c_j(A^*) = \begin{cases} \det A & \text{if } i = j, \\ 0 & \text{if } i \neq j. \end{cases}$$

Hence
$$A^{-1} = \frac{1}{\det A} A^*$$

so that if
$$A^{-1} = (c_{ij}), \text{ then } c_{ij} = \frac{A_{ji}}{\det A}.$$

We observe that this explicit form of the inverse is not practical for computation; and that A^{-1} does not exist if A is singular so that $\det A = 0$.

Let e^i denote the ith column of I. Thus if $n = 3$ then

$$e^1 = \begin{pmatrix} 1 \\ 0 \\ 0 \end{pmatrix} \qquad e^2 = \begin{pmatrix} 0 \\ 1 \\ 0 \end{pmatrix} \qquad e^3 = \begin{pmatrix} 0 \\ 0 \\ 1 \end{pmatrix}$$

Similarly let c^i denote the ith column of A^{-1}. Thus

$$c^1 = (c_{11}, c_{21}, \ldots, c_{n1})^T.$$

Then the relation

$$AA^{-1} = I$$

can be put in the form

$$A(c^1, c^2, \ldots, c^n) = (e^1, e^2, \ldots, e^n)$$

where A^{-1} is built up from its n columns $(c^1, c^2, \ldots, c^n)$ and I is built up from its n columns $(e^1, e^2, \ldots, e^n)$. This gives a practical way for computing the elements of the inverse matrix: we solve $Ax = b$, the system of equations (6.1.1), for $b = e^1$, then $e^2, \ldots$, then e^n giving $x = c^1$, then $c^2, \ldots$, then c^n respectively and A^{-1} is built up from these columns.

6.2 Solution of Sets of Linear Equations by Elimination

Throughout this section we take n, the number of unknowns, to be 4 as an illustration. It should be emphasized that the methods apply to any other value of n. We consider first two simple systems of equations:

$$\left.\begin{aligned}
x_1 \qquad\qquad\qquad\qquad &= b_1 \\
l_{21}x_1 + \quad x_2 \qquad\qquad &= b_2 \\
l_{31}x_1 + l_{32}x_2 + \quad x_3 \qquad &= b_3 \\
l_{41}x_1 + l_{42}x_2 + l_{43}x_3 + x_4 &= b_4,
\end{aligned}\right\} \qquad (6.2.1)$$

in matrix form

$$Lx = b,$$

where L, the matrix of the coefficients, is a lower triangular matrix with units as the diagonal elements. This system of equations can be solved simply thus:

$$\begin{aligned}
x_1 &= b_1 \\
x_2 &= b_2 - l_{21}x_1 \\
x_3 &= b_3 - l_{31}x_1 - l_{32}x_2 \\
x_4 &= b_4 - l_{41}x_1 - l_{42}x_2 - l_{43}x_3.
\end{aligned}$$

This procedure is called *forward substitution*.

Next we consider the system of equations:

$$\left.\begin{array}{r}
u_{11}x_1 + u_{12}x_2 + u_{13}x_3 + u_{14}x_4 = b_1 \\
u_{22}x_2 + u_{23}x_3 + u_{24}x_4 = b_2 \\
u_{33}x_3 + u_{34}x_4 = b_3 \\
u_{44}x_4 = b_4,
\end{array}\right\} \quad (6.2.2)$$

in matrix form

$$Ux = b,$$

where U is the upper triangular matrix of the coefficients. This system of equations is solved simply thus:

$$x_4 = b_4/u_{44}$$
$$x_3 = (b_3 - u_{34}x_4)/u_{33}$$
$$x_2 = (b_2 - u_{23}x_3 - u_{24}x_4)/u_{22}$$
$$x_1 = (b_1 - u_{12}x_2 - u_{13}x_3 - u_{14}x_4)/u_{11}.$$

This procedure is called *backward substitution*. If the same system of equations (6.2.2) is to be solved for different side columns b, then it may save time if we replace u_{ii} by their reciprocals $1/u_{ii}$ and multiply by this reciprocal instead of dividing by u_{ii} as multiplication is usually faster than division.

Next we consider the general system of equations:

$$\left.\begin{array}{r}
a_{11}x_1 + a_{12}x_2 + a_{13}x_3 + a_{14}x_4 = b_1 \\
a_{21}x_1 + a_{22}x_2 + a_{23}x_3 + a_{24}x_4 = b_2 \\
a_{31}x_1 + a_{32}x_2 + a_{33}x_3 + a_{34}x_4 = b_3 \\
a_{41}x_1 + a_{42}x_2 + a_{43}x_3 + a_{44}x_4 = b_4.
\end{array}\right\} \quad (6.2.3)$$

In this section we explain the method of elimination for solving linear equations in a way suitable for programming on digital computers. The method is very simple, in fact the first method one learns at school. We store the coefficients and the right-hand side constants in the form

$$\left.\begin{array}{ccccc}
a_{11} & a_{12} & a_{13} & a_{14} & b_1 \\
a_{21} & a_{22} & a_{23} & a_{24} & b_2 \\
a_{31} & a_{32} & a_{33} & a_{34} & b_3 \\
a_{41} & a_{42} & a_{43} & a_{44} & b_4.
\end{array}\right\} \quad (6.2.3)[1]$$

The first row is taken as *pivotal row* and we subtract multiples m_{21}, m_{31}, m_{41} of the pivotal row from the 2nd, 3rd, 4th rows respectively to eliminate the coefficients of x_1 and replace the eliminated coefficients a_{21}, a_{31}, a_{41} by the *multipliers* m_{21}, m_{31}, m_{41} so that the table now becomes

$$\left.\begin{matrix}
u_{11} & u_{12} & u_{13} & u_{14} & b_1 \\
m_{21} & a'_{22} & a'_{23} & a'_{24} & b'_2 \\
m_{31} & a'_{32} & a'_{33} & a'_{34} & b'_3 \\
m_{41} & a'_{42} & a'_{43} & a'_{44} & b'_4,
\end{matrix}\right\} \qquad (6.2.3)^2$$

where

$$u_{11}=a_{11} \qquad u_{12}=a_{12} \qquad u_{13}=a_{13} \qquad u_{14}=a_{14}$$
$$m_{21}=a_{21}/u_{11}$$
$$a'_{22}=a_{22}-m_{21}u_{12} \qquad a'_{23}=a_{23}-m_{21}u_{13}$$
$$a'_{24}=a_{24}-m_{21}u_{14} \qquad b'_2=b_2-m_{21}b_1$$
$$m_{31}=a_{31}/u_{11} \qquad \cdots$$
$$m_{41}=a_{41}/u_{11} \qquad \cdots$$

The leading coefficient $u_{11} = a_{11}$ is called the *pivot*, it occurs in the denominators of the expressions for the multipliers m_{21}, m_{31}, m_{41}. Now we repeat the procedure for the reduced system of equations

$$\left.\begin{matrix}
a'_{22}x_2 + a'_{23}x_3 + a'_{24}x_4 = b'_2 \\
a'_{32}x_2 + a'_{33}x_3 + a'_{34}x_4 = b'_3 \\
a'_{42}x_2 + a'_{43}x_3 + a'_{44}x_4 = b'_4
\end{matrix}\right\} \qquad (6.2.4)$$

and continue thus, reducing the dimension by unity at each step, to obtain

$$\left.\begin{matrix}
u_{11} & u_{12} & u_{13} & u_{14} & b_1 \\
m_{21} & u_{22} & u_{23} & u_{24} & b'_2 \\
m_{31} & m_{32} & a''_{33} & a''_{34} & b''_3 \\
m_{41} & m_{42} & a''_{43} & a''_{44} & b''_4
\end{matrix}\right\} \qquad (6.2.3)^3$$

$$\left.\begin{matrix}
u_{11} & u_{12} & u_{13} & u_{14} & b_1 \\
m_{21} & u_{22} & u_{23} & u_{24} & b'_2 \\
m_{31} & m_{32} & u_{33} & u_{34} & b''_3 \\
m_{41} & m_{42} & m_{43} & u_{44} & b'''_4.
\end{matrix}\right\} \qquad (6.2.3)^4$$

The reduced set of equations is now of the form $\mathbf{Ux} = \mathbf{b}'$ and is solved by backward substitution as in (6.2.2).

Remark 1. As the order in which the equations are written is immaterial, we can choose the pivotal row, when going from one stage to another, to be any of the equations in the set to be reduced. This could be done in a program either by a physical interchange of the rows so that the top row is the chosen pivotal row, or by keeping a record of the position of the row chosen as pivotal row. To explain the method of choice we take as an example the reduction from stage 2 to stage 3. Here we are eliminating x_2 in the set of equations (6.2.4). Normally one takes the first equation of this set to give the pivotal row so that

$$m_{32} = a'_{32}/a'_{22} \qquad m_{42} = a'_{42}/a'_{22}$$

Now a'_{22} was derived from the equation

$$a'_{22} = a_{22} - m_{21}u_{12};$$

it may happen that

$$a_{22} \sim m_{21}u_{12}$$

so that there is considerable cancellation and a large relative error. In an extreme case it may even happen that $a'_{22} = 0$ and it is then impossible to divide by a'_{22}. To avoid this rapid propagation of large errors we choose as a pivotal equation

$$u_{22}x_2 + u_{23}x_3 + u_{24}x_4 = b'_2,$$

that equation of the set (6.2.4) which has the numerically largest coefficient of x_2. For this choice of the pivot the multipliers m_{32}, m_{42} will be numerically less than 1. This modified form of elimination is called the method of *partial pivoting*; in this method all the multipliers are numerically less than 1, and this ensures that errors are not propagated by large multipliers. Under this method the order of the equations may be changed, but the unknowns are eliminated in the given order: first x_1 is eliminated, then x_2, ..., and so on. If we do not insist on eliminating the unknowns in this order, then we can choose as pivot at each stage the numerically largest coefficient of the whole matrix; this is called the method of *complete pivoting*. It is more complicated to program and usually it does not appreciably improve accuracy when compared with the method of partial pivoting.

Remark 2. The program could be arranged to solve the same set of equations for m different right-hand side vectors **b**, where m is a parameter to be specified. The same multipliers m_{ij} and the same coefficients u_{ij} are used for the different right-hand sides, and the elimination and backward substitution procedures could be carried out on each right-hand side, one at a time, or on all of them simultaneously.

Remark 3. By choosing the right-hand sides to be the columns of the unit matrix (e^1, e^2, e^3, e^4) the resulting solutions are the columns of the inverse matrix A^{-1}, as explained at the end of §6.1.

Example. To invert the matrix

$$A = \begin{pmatrix} 1 & 1 & 1 & 1 \\ 1 & 2 & 2 & 2 \\ 1 & 2 & 3 & 3 \\ 1 & 2 & 3 & 4 \end{pmatrix}$$

we solve the set of equations

$$\begin{aligned} x_1 + x_2 + x_3 + x_4 &= 1,\ 0,\ 0,\ 0 \\ x_1 + 2x_2 + 2x_3 + 2x_4 &= 0,\ 1,\ 0,\ 0 \\ x_1 + 2x_2 + 3x_3 + 3x_4 &= 0,\ 0,\ 1,\ 0 \\ x_1 + 2x_2 + 3x_3 + 4x_4 &= 0,\ 0,\ 0,\ 1. \end{aligned}$$

Stage 1

1	1	1	1	1	0	0	0
1	2	2	2	0	1	0	0
1	2	3	3	0	0	1	0
1	2	3	4	0	0	0	1

6

Stage 2

1	1	1	1	1	0	0	0
1	1	1	1	−1	1	0	0
1	1	2	2	−1	0	1	0
1	1	2	3	−1	0	0	1

Stage 3

1	1	1	1	1	0	0	0
1	1	1	1	−1	1	0	0
1	1	1	1	0	−1	1	0
1	1	1	2	0	−1	0	1

Stage 4

1	1	1	1	1	0	0	0
1	1	1	1	−1	1	0	0
1	1	1	1	0	−1	1	0
1	1	1	1	0	0	−1	1

We now solve the set of equations $\mathbf{U}\mathbf{x} = \mathbf{b}$ by back substitution, where

$$\mathbf{U} = \begin{pmatrix} 1 & 1 & 1 & 1 \\ 0 & 1 & 1 & 1 \\ 0 & 0 & 1 & 1 \\ 0 & 0 & 0 & 1 \end{pmatrix}.$$

We take

$$\mathbf{b} = \begin{pmatrix} 1 \\ -1 \\ 0 \\ 0 \end{pmatrix}, \begin{pmatrix} 0 \\ 1 \\ -1 \\ 0 \end{pmatrix}, \begin{pmatrix} 0 \\ 0 \\ 1 \\ -1 \end{pmatrix} \begin{pmatrix} 0 \\ 0 \\ 0 \\ 1 \end{pmatrix}$$

and obtain

$$\mathbf{x} = \begin{pmatrix} 2 \\ -1 \\ 0 \\ 0 \end{pmatrix}, \begin{pmatrix} -1 \\ 2 \\ -1 \\ 0 \end{pmatrix}, \begin{pmatrix} 0 \\ -1 \\ 2 \\ -1 \end{pmatrix}, \begin{pmatrix} 0 \\ 0 \\ -1 \\ 1 \end{pmatrix}$$

respectively.

Hence

$$\mathbf{A}^{-1} = \begin{pmatrix} 2 & -1 & 0 & 0 \\ -1 & 2 & -1 & 0 \\ 0 & -1 & 2 & -1 \\ 0 & 0 & -1 & 1 \end{pmatrix}.$$

Remark 4. To evaluate det $\mathbf{A}$, the determinant of a square matrix, we carry out the elimination exactly as in $(6.2.3)^1$ to $(6.2.3)^4$ without any right-hand sides. The determinant of the matrix of the coefficients (a_{ij}), with zeros in place of the multipliers m_{ij}, does not change from stage to stage. Thus

det $\mathbf{A}$ in stage 1 = det $\mathbf{U}$ in stage 4

$$= u_{11}u_{22}u_{33}u_{44}.$$

If the elimination is done with interchanges, which is safer and more accurate, we should multiply by -1 for every interchange.

Remark 5. We explain here an iterative method for improving the solution obtained by the method of elimination. But first we consider the practical computation of the scalar products of vectors; we recall the definition of $\mathbf{x} \cdot \mathbf{y}$:

$$\mathbf{x} \cdot \mathbf{y} = x_1 y_1 + x_2 y_2 + \ldots + x_n y_n.$$

It frequently happens that the vectors $\mathbf{x}$, $\mathbf{y}$ are nearly orthogonal so that considerable cancellation takes place when accumulating the products $x_i y_i$. The components x_i, y_i are usually single length numbers and the exact product of x_i and y_i is a double length number; we can form $\mathbf{x} \cdot \mathbf{y}$ much more accurately if we accumulate these double length products and then shift and round off the double length sum to single length. If, however, we round off each product to single length and then add, the result can be very inaccurate especially when $\mathbf{x}$ and $\mathbf{y}$ are nearly orthogonal. Given a matrix $\mathbf{A}$ and a vector $\mathbf{x}$, then each component of the vector $\mathbf{y} = \mathbf{A}\mathbf{x}$ is a scalar product; in what follows in this remark all such scalar products are formed by accumulating the double length products and then rounding to single length.

Let the vector $\mathbf{x}^{(1)}$ be the solution of the system $\mathbf{A}\mathbf{x} = \mathbf{b}$ obtained by the method of elimination, with or without pivoting. We compute the remainder vector $\mathbf{r}^{(1)}$ defined by the equation

$$\mathbf{r}^{(1)} = \mathbf{b} - \mathbf{A}\mathbf{x}^{(1)}.$$

If $\mathbf{x}^{(1)}$ were the exact solution, the remainder $\mathbf{r}^{(1)}$ would be zero, but of course this will not happen in practice. The solution may be unsatisfactory even though we use elimination with interchanges; this usually happens because much cancellation takes place in the calculation of the coefficients u_{ij}; this situation occurs frequently when n, the size of the set of equations, is large.

We distinguish two reasons why the solution may be unsatisfactory.

First, the solution may be unsatisfactory because some or all the components of the residual are not sufficiently small. To improve the solution in this sense it is necessary to carry out the elimination procedure using double length arithmetic or even higher precision. Such a system of equations is said to be ill-conditioned. It is beyond the scope of this book to give a more precise definition of the terms used here, in particular one needs to make more precise the expression "sufficiently small" which is relative and depends upon the order of magnitude of the elements in $\mathbf{A}$ and $\mathbf{b}$ and the required accuracy. But the general definition of an ill-conditioned problem is still useful here: *a small change in the data*, i.e. **A** or **b**, *causes a large change in the solution*.

Secondly, the solution may be unsatisfactory because $x^{(1)}$, the solution obtained, differs considerably from the true solution, for example it may have only one or two correct figures. It may be possible to improve the accuracy of the solution in this sense, without having to repeat most of the elimination work and without having to use double length arithmetic except when accumulating products as explained above. We explain now a method for doing this. But we emphasize that this method may improve the accuracy in the solution, but it will not usually make the residual smaller unless the initial approximate solution was very inaccurate or derived by some method other than elimination.

We solve the set of equations

$$Ax = r^{(1)},$$

giving
$$x = c^{(1)}.$$

The solution is obtained readily as the multipliers m_{ij} and the coefficients u_{ij} do not have to be computed again. The solution $c^{(1)}$ is said to be the correction vector and

$$x^{(2)} = x^{(1)} + c^{(1)}$$

is usually a better solution than $x^{(1)}$. We can repeat this procedure iteratively: we compute

$$r^{(i)} = b - Ax^{(i)},$$

$$Ax = r^{(i)} \text{ gives } x = c^{(i)},$$

$$x^{(i+1)} = x^{(i)} + c^{(i)}.$$

We stop when the components of the correction vector are sufficiently small. This procedure usually improves the solution without reducing the residual.

Remark 6. In many cases the multipliers m_{ij} need not be stored as they are not needed in later calculations; but in a general library program it is better to preserve the multipliers as this takes up no extra space. For example the method of the previous remark for improving the solution would not be practical if the multipliers m_{ij} were not recorded for later use.

6.3 Solution of Sets of Linear Equations on Desk Machines

The method of this section is essentially the same as the method of elimination of the previous section, but it is stated in a form suitable for desk machines because it does not require intermediate recordings on the computing sheet. However, many of the ideas introduced in this section are also relevant to digital computers. For simplicity we shall not take interchanges into account though this could be done with a little care. As before we take the case $n = 4$ for illustration.

Given a system of equations $\mathbf{Ax} = \mathbf{b}$, we find a lower triangular matrix $\mathbf{L}$ of the form (6.2.1) and an upper triangular matrix $\mathbf{U}$ of the form (6.2.2) such that

$$\mathbf{A} = \mathbf{LU}.$$

Then $\mathbf{Ax} = \mathbf{b}$ takes the form

$$\mathbf{LUx} = \mathbf{b}.$$

If we set $\mathbf{Ux} = \mathbf{y}$, then the system $\mathbf{Ax} = \mathbf{b}$ becomes $\mathbf{Ly} = \mathbf{b}$. This can be solved for $\mathbf{y}$ by forward substitution as in (6.2.1). Then we can solve $\mathbf{Ux} = \mathbf{y}$ for $\mathbf{x}$ by backward substitution as in (6.2.2).

Example. Let

$$\mathbf{A} = \begin{pmatrix} 1 & 1 & 1 & 1 \\ 1 & 2 & 2 & 2 \\ 1 & 2 & 3 & 3 \\ 1 & 2 & 3 & 4 \end{pmatrix}, \quad \mathbf{L} = \begin{pmatrix} 1 & 0 & 0 & 0 \\ 1 & 1 & 0 & 0 \\ 1 & 1 & 1 & 0 \\ 1 & 1 & 1 & 1 \end{pmatrix}, \quad \mathbf{U} = \begin{pmatrix} 1 & 1 & 1 & 1 \\ 0 & 1 & 1 & 1 \\ 0 & 0 & 1 & 1 \\ 0 & 0 & 0 & 1 \end{pmatrix}$$

We can verify that $\mathbf{LU} = \mathbf{A}$. To solve $\mathbf{Ax} = \mathbf{b}$ where

$$\mathbf{b} = (0, \ 1, \ 1, \ 0)^{\mathrm{T}}$$

we first solve the system $\mathbf{Ly} = \mathbf{b}$:

$$
\begin{aligned}
y_1 \qquad\qquad\qquad &= 0 \\
y_1 + y_2 \qquad\qquad &= 1 \\
y_1 + y_2 + y_3 \qquad &= 1 \\
y_1 + y_2 + y_3 + y_4 &= 0
\end{aligned}
$$

which gives $\mathbf{y} = \begin{pmatrix} 0 \\ 1 \\ 0 \\ -1 \end{pmatrix}$

Next we solve the system $\mathbf{Ux} = \mathbf{y}$:

$$
\begin{aligned}
x_1 + x_2 + x_3 + x_4 &= 0 \\
x_2 + x_3 + x_4 &= 1 \\
x_3 + x_4 &= 0 \\
x_4 &= -1
\end{aligned}
$$

which gives $\mathbf{x} = \begin{pmatrix} -1 \\ 1 \\ 1 \\ -1 \end{pmatrix}$.

Now we explain how to calculate the elements l_{ij}, u_{ij} of $\mathbf{L}$ and $\mathbf{U}$. Spaces on the computation sheet are laid out in the form

$$
\begin{matrix}
u_{11} & u_{12} & u_{13} & u_{14} \\
l_{21} & u_{22} & u_{23} & u_{24} \\
l_{31} & l_{32} & u_{33} & u_{34} \\
l_{41} & l_{42} & l_{43} & l_{44}
\end{matrix}
$$

The elements l_{ij}, u_{ij} are calculated in the order in which they are laid out, viz.,

$$u_{11}, \ u_{12}, \ u_{13}, \ u_{14}, \ l_{21}, \ u_{22}, \ u_{23}, \ u_{24}, \ l_{31}, \ l_{32}, \ u_{33}, \ \ldots$$

From $\mathbf{A} = \mathbf{LU}$ we have:

For $j < i$

$$a_{ij} = l_{i1}u_{1j} + l_{i2}u_{2j} + \ldots + l_{i, \ j-1}u_{j-1, j} + l_{ij}u_{jj};$$

this is solved for l_{ij}, for $j = 1, 2, 3, \ldots, i-1$:

$$l_{ij} = (a_{ij} - l_{i1}u_{1j} - l_{i2}u_{2j} - \ldots - l_{i,\,j-1}u_{j-1,j})/u_{jj}.$$

We observe that according to the order outlined above all the elements on the right-hand of this expression will have been computed when we reach the position (ij); we also observe that this rather complicated expression for l_{ij} can be computed on a desk machine without any intermediate recordings by accumulating the products and dividing into the accumulated sum.

For $j \geqq i$

$$a_{ij} = l_{i1}u_{1j} + l_{i2}u_{2j} + \ldots + l_{i,i-1}u_{i-1,j} + u_{ij};$$

this is solved for u_{ij} for $j = i, i+1, \ldots, n$:

$$u_{ij} = a_{ij} - l_{i1}u_{1j} - l_{i2}u_{2j} - \ldots - l_{i,i-1}u_{i-1,j}.$$

The same observations apply here as for l_{ij}.

Example. If $n = 4$, the computation is as follows:
For $i = 1$ and $j = 1, 2, 3, 4$ we have

$$u_{11} = a_{11}, \quad u_{12} = a_{12}, \quad u_{13} = a_{13}, \quad u_{14} = a_{14}.$$

For $i = 2$ and $j = 1$ we have

$$l_{21} = a_{21}/u_{11}.$$

For $i = 2$ and $j = 2, 3, 4$ we have

$$u_{22} = a_{22} - l_{21}u_{12}, \quad u_{23} = a_{23} - l_{21}u_{13}, \quad u_{24} = a_{24} - l_{21}u_{14}.$$

For $i = 3$ and $j = 1, 2$ we have

$$l_{31} = a_{31}/u_{11}, \quad l_{32} = (a_{32} - l_{31}u_{12})/u_{22}.$$

For $i = 3$ and $j = 3, 4$ we have

$$u_{33} = a_{33} - l_{31}u_{13} - l_{32}u_{23}, \quad u_{34} = a_{34} - l_{31}u_{14} - l_{32}u_{24}.$$

For $i = 4$ and $j = 1, 2, 3$ we have

$$l_{41} = a_{44}/u_{11},$$
$$l_{42} = (a_{42} - l_{41}u_{12})/u_{22},$$
$$l_{43} = (a_{43} - l_{41}u_{13} - l_{42}u_{23})/u_{33}.$$

For $i = 4$ and $j = 4$ we have

$$u_{44} = a_{44} - l_{41}u_{14} - l_{42}u_{24} - l_{43}u_{34}.$$

Remark 1. For general i, j, n the expression for u_{ij} consists of the sum of products and can be put in the form of a scalar product of two vectors; the same remark applies to the numerator in the expression for l_{ij}. We observe that the advantage of this method for desk machines is that such expressions are easy to compute without intermediate recordings. This method is also very suitable for digital computers which have facilities for the accumulation of double length products of single length numbers in the manner explained at the beginning of Remark 5 of the previous section. Elimination programmed in this way will therefore be more accurate than the way explained in the previous section. However programming this method is a little harder if we do the elimination with interchanges or if the digital computer has a small immediate access store.

Remark 2. The solution of $\mathbf{Ax} = \mathbf{b}$ could be incorporated in the computation scheme for $\mathbf{L}$ and $\mathbf{U}$. We also introduce check columns which are important in any desk computation.

We start with the following table

$$\left.\begin{array}{cccccc} a_{11} & a_{12} & a_{13} & a_{14} & b_1 & c_1 \\ a_{21} & a_{22} & a_{23} & a_{24} & b_2 & c_2 \\ a_{31} & a_{32} & a_{33} & a_{34} & b_3 & c_3 \\ a_{41} & a_{42} & a_{43} & a_{44} & b_4 & c_4 \end{array}\right\}, \qquad (6.3.1)$$

where the c column is the sum of the other columns, e.g.

$$c_2 = a_{21} + a_{22} + a_{23} + a_{24} + b_2$$

Next we calculate the elements in the following table

$$\left.\begin{array}{cccccc} u_{11} & u_{12} & u_{13} & u_{14} & y_1 & d_1 \\ l_{21} & u_{22} & u_{23} & u_{24} & y_2 & d_2 \\ l_{31} & l_{32} & u_{33} & u_{34} & y_3 & d_3 \\ l_{41} & l_{42} & l_{43} & u_{44} & y_4 & d_4 \end{array}\right\} \qquad (6.3.2)$$

The elements are calculated in the order in which they are written. The l's and u's are calculated as explained before. The y's and d's are calculated like the u's from the relations $\mathbf{Ly} = \mathbf{b}$, $\mathbf{Ld} = \mathbf{c}$ as follows:

for $i = 1$
$$y_1 = b_1, \quad d_1 = c_1$$
for $i = 2$
$$y_2 = b_2 - l_{21}y_1, \quad d_2 = c_2 - l_{21}d_1$$
for $i = 3$
$$y_3 = b_3 - l_{31}y_1 - l_{32}y_2, \quad d_3 = c_3 - l_{31}d_1 - l_{32}d_2$$
for $i = 4$
$$y_4 = b_4 - l_{41}y_1 - l_{42}y_2 - l_{43}y_3, \quad d_4 = c_4 - l_{41}d_1 - l_{42}d_2 - l_{43}d_3.$$

As each row of (6.3.2) is calculated we check that the corresponding d is equal to the sum of the remaining elements in that row before proceeding to calculate the next row; if there is a difference of a few units in the least significant digit this can be explained by rounding errors and we replace the computed d by the exact sum, otherwise the calculation for that row has to be repeated. The calculation of the first row in (6.3.2) is trivial as it is the same as the first row of (6.3.1). When (6.3.2) is completed we find $\mathbf{x}$ by solving $\mathbf{Ux} = \mathbf{y}$ by backward substitution. As a final check we could work out f_1, f_2, f_3, f_4, f_5 the sums of the columns in (6.3.1) and check that

$$f_1x_1 + f_2x_2 + f_3x_3 + f_4x_4 = f_5.$$

Remark 3. A little study of the way l_{ij}, u_{ij} are derived in this section will show that they are the same as the m_{ij}, u_{ij} of the previous section. We now demonstrate this formally. Let $\mathbf{A}^{(1)}$ be the matrix of the coefficients in $(6.2.3)^1$,

$$\mathbf{A}^{(1)} = \begin{pmatrix} a_{11} & a_{12} & a_{13} & a_{14} \\ a_{21} & a_{22} & a_{23} & a_{24} \\ a_{31} & a_{32} & a_{33} & a_{34} \\ a_{41} & a_{42} & a_{43} & a_{44} \end{pmatrix}.$$

Let $\mathbf{A}^{(2)}$ be the matrix of the coefficients in $(6.2.3)^2$ so that there are zeros instead of the multipliers:

$$\mathbf{A}^{(2)} = \begin{pmatrix} u_{11} & u_{12} & u_{13} & u_{14} \\ 0 & a'_{22} & a'_{23} & a'_{24} \\ 0 & a'_{32} & a'_{33} & a'_{34} \\ 0 & a'_{42} & a'_{43} & a'_{44} \end{pmatrix}.$$

Similarly,

$$\mathbf{A}^{(3)} = \begin{pmatrix} u_{11} & u_{12} & u_{13} & u_{14} \\ 0 & u_{22} & u_{23} & u_{24} \\ 0 & 0 & a''_{33} & a''_{34} \\ 0 & 0 & a''_{43} & a''_{44} \end{pmatrix}.$$

and

$$\mathbf{A}^{(4)} = \begin{pmatrix} u_{11} & u_{12} & u_{13} & u_{14} \\ 0 & u_{22} & u_{23} & u_{24} \\ 0 & 0 & u_{33} & u_{34} \\ 0 & 0 & 0 & u_{44} \end{pmatrix}.$$

Let

$$\mathbf{M}_1 = \begin{pmatrix} 1 & 0 & 0 & 0 \\ -m_{21} & 1 & 0 & 0 \\ -m_{31} & 0 & 1 & 0 \\ -m_{41} & 0 & 0 & 1 \end{pmatrix}, \quad \mathbf{M}_2 = \begin{pmatrix} 1 & 0 & 0 & 0 \\ 0 & 1 & 0 & 0 \\ 0 & -m_{32} & 1 & 0 \\ 0 & -m_{42} & 0 & 1 \end{pmatrix},$$

$$\mathbf{M_3} = \begin{pmatrix} 1 & 0 & 0 & 0 \\ 0 & 1 & 0 & 0 \\ 0 & 0 & 1 & 0 \\ 0 & 0 & -m_{43} & 0 \end{pmatrix}, \quad \mathbf{L} = \begin{pmatrix} 1 & 0 & 0 & 0 \\ m_{21} & 1 & 0 & 0 \\ m_{31} & m_{32} & 1 & 0 \\ m_{41} & m_{42} & m_{43} & 1 \end{pmatrix}$$

Then we can verify that

$$\mathbf{M_1^{-1}} = \begin{pmatrix} 1 & 0 & 0 & 0 \\ m_{21} & 1 & 0 & 0 \\ m_{31} & 0 & 1 & 0 \\ m_{41} & 0 & 0 & 1 \end{pmatrix},$$

$$\mathbf{M_2^{-1}} = \begin{pmatrix} 1 & 0 & 0 & 0 \\ 0 & 1 & 0 & 0 \\ 0 & m_{32} & 1 & 0 \\ 0 & m_{42} & 0 & 1 \end{pmatrix}, \quad \mathbf{M_3^{-1}} = \begin{pmatrix} 1 & 0 & 0 & 0 \\ 0 & 1 & 0 & 0 \\ 0 & 0 & 1 & 0 \\ 0 & 0 & m_{43} & 1 \end{pmatrix}$$

$$\mathbf{L} = \mathbf{M_1^{-1}M_2^{-1}M_3^{-1}}$$

$$\mathbf{A}^{(2)} = \mathbf{M_1A}^{(1)} \quad \mathbf{A}^{(3)} = \mathbf{M_2A}^{(2)} \quad \mathbf{U} = \mathbf{A}^{(4)} = \mathbf{M_3A}^{(3)} = \mathbf{M_3M_2M_1A}$$

$$\mathbf{A} = \mathbf{M_1^{-1}\,M_2^{-1}\,M_3^{-1}\,U} = \mathbf{LU}.$$

It can also be shown that we cannot have two different triangular decompositions of $\mathbf{A}$ in the form $\mathbf{LU}$ where $\mathbf{L}$ has units along the diagonal. Hence the multipliers m_{ij} of the previous section are the same as the l_{ij} of this section; the only difference between the two methods is the order in which the elements of $\mathbf{L}$ and $\mathbf{U}$ are calculated.

Remark 4. Matrices are often inverted unnecessarily. If we calculate the matrices L and U so that $LU = A$ then for any vector **b** we can calculate $x = A^{-1}b$ by first solving $Ly = b$ for **y** by forward substitution and then $Ux = y$ by backward substitution. The resulting **x** is obtained by n^2 multiplications which is exactly the number of multiplications necessary for calculating $A^{-1}b$ by direct multiplication. The storage for L and U is the same as that required for A^{-1} and much more work is necessary to find A^{-1} than to find L and U. (See Example 2 in the next section.)

6.4 Latent Roots and Vectors

The terms *characteristic roots* and *characteristic vectors* or *eigenroots* and *eigenvectors* are sometimes used instead of latent roots and latent vectors. The subject is important because it turns up in many branches of the pure and applied sciences and in statistics. A full treatment is beyond the scope of this book and this section contains only a very brief introduction, starting with a few more results from the general theory of vectors and matrices which we need here.

Given an $(n \times n)$ matrix A and a vector **x** then A**x** is itself a vector which can be thought of as a combination of the column vectors of A. Thus if

$$x = (x_1, x_2, \ldots, x_n)^T$$

and the column vectors of A are denoted by

$$C^{(1)}, C^{(2)}, \ldots, C^{(n)},$$

then

$$Ax = x_1 C^{(1)} + x_2 C^{(2)} + \ldots + x_n C^{(n)}.$$

It follows that if $Ax = 0$ for a non-zero vector **x** then the columns of A are linearly dependent and hence det $A = 0$. Conversely if det $A = 0$ then we can find at least one non-zero vector **x** such that $Ax = 0$.

We recall that two real vectors **x**, **y** are said to be orthogonal if

$\mathbf{x} \cdot \mathbf{y} = 0$. The component of $\mathbf{x}$ along $\mathbf{y}$ is

$$k\mathbf{y} \text{ where } k = (\mathbf{x} \cdot \mathbf{y})/(\mathbf{y} \cdot \mathbf{y})$$

If we extract from $\mathbf{x}$ its component along $\mathbf{y}$ we obtain a new vector $\mathbf{x}'$ which is orthogonal to $\mathbf{y}$:

$$\mathbf{x}' = \mathbf{x} - k\mathbf{y} \qquad \mathbf{x}' \cdot \mathbf{y} = 0$$

Given n linearly independent vectors of dimension n

$$\mathbf{y}^{(1)}, \mathbf{y}^{(2)}, \ldots, \mathbf{y}^{(n)},$$

then any n-dimensional non-zero vector $\mathbf{x}$ can be expressed as a linear combination of $\mathbf{y}^{(1)}, \mathbf{y}^{(2)}, \ldots, \mathbf{y}^{(n)}$

$$\mathbf{x} = k_1 \mathbf{y}^{(1)} + k_2 \mathbf{y}^{(2)} + \ldots + k_n \mathbf{y}^{(n)},$$

where $k_1, k_2, \ldots, k_n$ are scalars. We say that the set of vectors $\mathbf{y}^{(1)}, \mathbf{y}^{(2)}, \ldots, \mathbf{y}^{(n)}$ form a frame of reference so that any other vector can be expressed in terms of them. The set is said to be orthogonal if $\mathbf{y}^{(i)} \cdot \mathbf{y}^{(j)} = 0$ when $i \neq j$; then the constants k_i in the expression above are given by the equations

$$k_i = (\mathbf{x} \cdot \mathbf{y}^{(i)})/(\mathbf{y}^{(i)} \cdot \mathbf{y}^{(i)})$$

derived by forming the scalar product of the two sides of the expression with $\mathbf{y}^{(i)}$.

A vector $\mathbf{x}$ is said to be a *latent vector* of a matrix $\mathbf{A}$ if

$$\mathbf{A}\mathbf{x} = \lambda\mathbf{x}$$

where λ is a scalar.

Example. Let

$$\mathbf{A} = \begin{pmatrix} 5 & 2 \\ 2 & 2 \end{pmatrix}$$

then $\mathbf{x} = \begin{pmatrix} 2 \\ 1 \end{pmatrix}$ is a latent vector of $\mathbf{A}$ since

$$\mathbf{A}\mathbf{x} = \begin{pmatrix} 12 \\ 6 \end{pmatrix} = 6\mathbf{x}.$$

The scalar quantity λ is said to be the *latent root* of $\mathbf{A}$ corresponding to the latent vector $\mathbf{x}$. The relation $\mathbf{Ax} = \lambda\mathbf{x}$ can be put in the form

$$\mathbf{Ax} = \lambda\mathbf{Ix} \text{ hence } (\mathbf{A} - \lambda\mathbf{I})\,\mathbf{x} = 0,$$

where $\mathbf{I}$ is the unit matrix. Thus in the above example

$$\mathbf{A} - \lambda\mathbf{I} = \begin{pmatrix} 5 & 2 \\ 2 & 2 \end{pmatrix} - 6\begin{pmatrix} 1 & 0 \\ 0 & 1 \end{pmatrix} = \begin{pmatrix} -1 & 2 \\ 2 & -4 \end{pmatrix}$$

and $\quad (\mathbf{A} - \lambda\mathbf{I})\,\mathbf{x} = \begin{pmatrix} -1 & 2 \\ 2 & -4 \end{pmatrix}\begin{pmatrix} 2 \\ 1 \end{pmatrix} = \begin{pmatrix} 0 \\ 0 \end{pmatrix}$

It follows that

$$\det(\mathbf{A} - \lambda\mathbf{I}) = 0.$$

When this is expanded for a general determinant of order n we get a polynomial of order n in λ which is said to be the *characteristic polynomial*. Thus there are in general n latent roots and n corresponding latent vectors. In our example

$$\mathbf{A} - \lambda\mathbf{I} = \begin{pmatrix} 5 - \lambda & 2 \\ 2 & 2 - \lambda \end{pmatrix},$$

$$\det(\mathbf{A} - \lambda\mathbf{I}) = (5 - \lambda)(2 - \lambda) - 4 = \lambda^2 - 7\lambda + 6$$

which is zero when $\lambda = 6$ or 1. When $\lambda = 6$ we have

$$(\mathbf{A} - 6\mathbf{I})\,\mathbf{x} = \begin{pmatrix} -1 & 2 \\ 2 & -4 \end{pmatrix}\begin{pmatrix} x_1 \\ x_2 \end{pmatrix} = \begin{pmatrix} -x_1 + 2x_2 \\ 2x_1 - 4x_2 \end{pmatrix} = \begin{pmatrix} 0 \\ 0 \end{pmatrix}$$

which gives $\mathbf{x} = \begin{pmatrix} 2 \\ 1 \end{pmatrix}$ or any multiple of $\begin{pmatrix} 2 \\ 1 \end{pmatrix}$.

Thus the latent vector is determined up to an arbitrary multiplier; only the ratios of the components to each other are determined.

The latent vector corresponding to $\lambda = 6$ could be taken as

$$\begin{pmatrix} 2 \\ 1 \end{pmatrix} \text{ or } \begin{pmatrix} 4 \\ 2 \end{pmatrix} \text{ or } \begin{pmatrix} -2 \\ -1 \end{pmatrix}.$$

If we take $\lambda = 1$ we get,

$$(\mathbf{A} - \lambda\mathbf{I})\,\mathbf{x} = \begin{pmatrix} 4 & 2 \\ 2 & 1 \end{pmatrix} \begin{pmatrix} x_1 \\ x_2 \end{pmatrix} = \begin{pmatrix} 4x_1 + 2x_2 \\ 2x_1 + x_2 \end{pmatrix} = \begin{pmatrix} 0 \\ 0 \end{pmatrix}$$

which gives $\mathbf{x} = \begin{pmatrix} -1 \\ 2 \end{pmatrix}$.

In general let

$$\mathbf{A} = (a_{ij}),$$

$$\pm \det (\mathbf{A} - \lambda\mathbf{I}) = \lambda^n - p_1\lambda^{n-1} + p_2\lambda^{n-2} - \ldots \pm p_n.$$

It is obvious that

$$p_1 = a_{11} + a_{22} + \ldots + a_{nn}.$$

This expression, which is the sum of the diagonal elements of $\mathbf{A}$ is called the *trace* of $\mathbf{A}$ and denoted by tr $(\mathbf{A})$. If the latent roots are

$$\lambda_1, \lambda_2, \lambda_3, \ldots, \lambda_n,$$

it follows from the elementary properties of the roots of polynomials that

$$\text{tr }(\mathbf{A}) = \lambda_1 + \lambda_2 + \ldots + \lambda_n = \text{sum of the latent roots,}$$

and that

$$p_n = \det \mathbf{A} = \lambda_1\,\lambda_2\,\lambda_3\,\ldots\,\lambda_n$$

$$= \text{product of the latent roots.}$$

The first result provides a useful check in numerical work. In the above example

$$\text{tr }(\mathbf{A}) = 5 + 2, \ \lambda_1 + \lambda_2 = 6 + 1,$$
$$\det \mathbf{A} = 6, \ \lambda_1\lambda_2 = 6.$$

Very frequently in applications $\mathbf{A}$ is symmetric, i.e. $\mathbf{A}^\mathrm{T} = \mathbf{A}$ or $a_{ij} = a_{ji}$ so that for any two vectors $\mathbf{x}$, $\mathbf{y}$ we have

$$\mathbf{x} \cdot \mathbf{A}\mathbf{y} = \mathbf{y} \cdot \mathbf{A}\mathbf{x}.$$

In general, even though $\mathbf{A}$ may be real, some of its latent roots may be complex, however, if $\mathbf{A}$ is real and symmetric then all the latent roots are real. To see this, let λ be a latent root of $\mathbf{A}$ and $\mathbf{x}$ the corresponding latent vector. Let $\bar{\mathbf{x}}$ denote the vector whose components are the complex conjugates of the corresponding components of $\mathbf{x}$; we define $\bar{\mathbf{A}}$ similarly, because $\mathbf{A}$ is real we have $\bar{\mathbf{A}} = \mathbf{A}$. We have

$$\mathbf{A}\mathbf{x} = \lambda\mathbf{x},$$

hence $\qquad\qquad \bar{\mathbf{x}} \cdot \mathbf{A}\mathbf{x} = \lambda\bar{\mathbf{x}} \cdot \mathbf{x}.$

The complex conjugate of $\bar{\mathbf{x}} \cdot \mathbf{A}\mathbf{x}$ is

$$\mathbf{x} \cdot \bar{\mathbf{A}\mathbf{x}} = \mathbf{x} \cdot \mathbf{A}\bar{\mathbf{x}} = \bar{\mathbf{x}} \cdot \mathbf{A}\mathbf{x}$$

because of the symmetry of $\mathbf{A}$, hence $\bar{\mathbf{x}} \cdot \mathbf{A}\mathbf{x}$ must be real and $\bar{\mathbf{x}} \cdot \mathbf{x}$ is also real. It follows that

$$\lambda = \bar{\mathbf{x}} \cdot \mathbf{A}\mathbf{x}/(\bar{\mathbf{x}} \cdot \mathbf{x})$$

must also be real.

Another important property of symmetric matrices relates to the latent vectors. If λ_1, λ_2 are two distinct latent roots of the real symmetric matrix $\mathbf{A}$, and $\mathbf{x}$, $\mathbf{y}$ are the corresponding latent vectors then $\mathbf{x}$ and $\mathbf{y}$ are orthogonal, i.e. $\mathbf{x} \cdot \mathbf{y} = 0$. The proof is as follows:

from $\qquad\qquad \mathbf{A}\mathbf{x} = \lambda_1\mathbf{x}$ and $\mathbf{A}\mathbf{y} = \lambda_2\mathbf{y}$

we have $\qquad \mathbf{y} \cdot \mathbf{A}\mathbf{x} = \lambda_1\mathbf{y} \cdot \mathbf{x}$ and $\mathbf{x} \cdot \mathbf{A}\mathbf{y} = \lambda_2\mathbf{x} \cdot \mathbf{y},$

but $\qquad\qquad \mathbf{y} \cdot \mathbf{x} = \mathbf{x} \cdot \mathbf{y}$ and $\mathbf{y} \cdot \mathbf{A}\mathbf{x} = \mathbf{x} \cdot \mathbf{A}\mathbf{y}$

because of the symmetry of $\mathbf{A}$. Hence

$$\lambda_1\mathbf{x} \cdot \mathbf{y} = \lambda_2\mathbf{x} \cdot \mathbf{y}$$

which is only possible if $\mathbf{x} \cdot \mathbf{y} = 0$ since $\lambda_1 \neq \lambda_2$.

In general a real symmetric matrix of order n has n distinct

latent roots
$$\lambda_1, \lambda_2, \lambda_3, \ldots, \lambda_n$$

and n corresponding latent vectors

$$\mathbf{x}^{(1)}, \mathbf{x}^{(2)}, \mathbf{x}^{(3)}, \ldots, \mathbf{x}^{(n)}$$

which are mutually orthogonal. They must be therefore linearly independent, for otherwise there would be a linear relation of the form
$$c_1\mathbf{x}^{(1)} + c_2\mathbf{x}^{(2)} + \ldots + c_n\mathbf{x}^{(n)} = 0,$$

where at least one coefficient, say c_r, is not zero. Forming the scalar product of this relation with x_r we get $c_r\mathbf{x}^{(r)} . \mathbf{x}^{(r)} = 0$ but $\mathbf{x}^{(r)} . \mathbf{x}^{(r)} \neq 0$, hence $c_r = 0$ which is a contradiction. It follows that the n latent vectors form an orthogonal frame of reference so that any vector $\mathbf{x}$ can be expressed in the form

$$k_1\mathbf{x}^{(1)} + k_2\mathbf{x}^{(2)} + \ldots + k_n\mathbf{x}^{(n)}.$$

For real unsymmetric matrices, some of the latent roots are usually complex, but provided they are distinct we have still n linearly independent, though not orthogonal, latent vectors $\mathbf{x}^{(1)}, \mathbf{x}^{(2)}, \ldots, \mathbf{x}^{(n)}$.

It is sometimes possible to obtain an approximation $\mathbf{y}$ to the true latent vector $\mathbf{x}$ of a real symmetric matrix $\mathbf{A}$. Then it is possible to obtain a good approximation to the corresponding latent root from the *Rayleigh–Ritz* expression:

$$\lambda \sim \mathbf{y} . \mathbf{Ay}/(\mathbf{y} . \mathbf{y}).$$

For suppose that $\mathbf{y}$ is scaled so that $\mathbf{y} . \mathbf{y} = 1$ and that

$$\mathbf{y} = \mathbf{x} + \mathbf{e}$$

where $\mathbf{x}$ is the true latent vector and $\mathbf{e}$ is the error vector whose components may be regarded as infinitesimal of the first order and chosen so that $\mathbf{e}$ has no components along $\mathbf{x}$:

$$\mathbf{e} . \mathbf{x} = 0,$$

then
$$\mathbf{y} . \mathbf{Ay}/(\mathbf{y} . \mathbf{y}) = \mathbf{y} . \mathbf{Ay} = (\mathbf{x} + \mathbf{e}) . \mathbf{A}(\mathbf{x} + \mathbf{e})$$
$$= \mathbf{x} . \mathbf{Ax} + \mathbf{x} . \mathbf{Ae} + \mathbf{e} . \mathbf{Ax} + \mathbf{e} . \mathbf{Ae}.$$

Now by the symmetry of A we have

$$\mathbf{x} . \mathbf{Ae} = \mathbf{e} . \mathbf{Ax} = \lambda \mathbf{e} . \mathbf{x} = 0.$$

Furthermore

$$\mathbf{x} . \mathbf{Ax} = \lambda \mathbf{x} . \mathbf{x} \text{ and } \mathbf{x} . \mathbf{x} = \mathbf{y} . \mathbf{y} - \mathbf{e} . \mathbf{e} = 1 - \mathbf{e} . \mathbf{e}.$$

Hence

$$\mathbf{y} . \mathbf{Ay}/(\mathbf{y} . \mathbf{y}) = \lambda + \mathbf{e} . \mathbf{Ae} - \mathbf{e} . \mathbf{e},$$
$$= \lambda + \text{infinitesimal terms of the second order.}$$

We observe that if A is not symmetric then $\mathbf{x} . \mathbf{Ae} \neq \mathbf{e} . \mathbf{Ax}$ and then

$$\mathbf{y} . \mathbf{Ay}/(\mathbf{y} . \mathbf{y}) = \lambda + \text{infinitesimal terms of the first order.}$$

If $\mathbf{Ax} = \lambda \mathbf{x}$ then

$$\mathbf{A}^2\mathbf{x} = \mathbf{AAx} = \lambda \mathbf{Ax} = \lambda^2\mathbf{x}$$

and generally
$$\mathbf{A}^n\mathbf{x} = \lambda^n\mathbf{x},$$

so that if λ is a root of A then λ^2 is a latent root of $\mathbf{A}^2$ and $1/\lambda$ is a root of $\mathbf{A}^{-1}$. Similarly we can show that if λ is a latent root of A then $\lambda - h$ is a latent root of $\mathbf{A} - h\mathbf{I}$.

In this section we explain only one method, the iterative method, for finding the latent roots and vectors of matrices. The method is simple and capable of various extensions and improvements as will be seen; the ideas underlying it are basic in many other situations in numerical analysis. However, there are many more advanced and more efficient methods which are beyond the scope of this book.

Suppose that the latent roots and corresponding latent vectors of A are

$$\lambda_1, \lambda_2, \ldots, \lambda_n,$$

$$\mathbf{x}^{(1)}, \mathbf{x}^{(2)}, \ldots, \mathbf{x}^{(n)},$$

ordered so that

$$|\lambda_1| > |\lambda_2| \geq \ldots \geq |\lambda_n|.$$

We start with an arbitrary vector $\mathbf{y}$, say

$$\mathbf{y} = (1, 1, \ldots, 1)^\mathrm{T}.$$

Let

$$\mathbf{y} = k_1 \mathbf{x}^{(1)} + k_2 \mathbf{x}^{(2)} + \ldots + k_n \mathbf{x}^{(n)}.$$

We calculate iteratively

$$\mathbf{y}^{(1)} = \mathbf{A}\mathbf{y} = k_1 \lambda_1 \mathbf{x}^{(1)} + k_2 \lambda_2 \mathbf{x}^{(2)} + \ldots + k_n \lambda_n \mathbf{x}^{(n)},$$

$$y^{(2)} = \mathbf{A}\mathbf{y}^{(1)} = k_1 \lambda_1^2 \mathbf{x}^{(1)} + k_2 \lambda_2^2 \mathbf{x}^{(2)} + \ldots + k_n \lambda_n^2 \mathbf{x}^{(n)},$$

and generally

$$y^{(m)} = \mathbf{A}\mathbf{y}^{(m-1)} = k_1 \lambda_1^m \mathbf{x}^{(1)} + k_2 \lambda_2^m \mathbf{x}^{(2)} + \ldots + k_n \lambda_n^m \mathbf{x}^{(n)}.$$

Now when m is large enough

$$| \lambda_1^m | \gg | \lambda_2^m | \geqq \ldots \geqq | \lambda_n^m |$$

(where $\gg$ means "much greater than"), hence

$$\mathbf{y}^{(m)} \sim k_1 \lambda_1^m \mathbf{x}^{(1)}$$

and

$$\mathbf{y}^{(m+1)} \sim \lambda_1 \mathbf{y}^{(m)}.$$

If $| \lambda_1 | > 1$ the components of $\mathbf{y}^{(m)}$ get larger and larger as m increases; if $| \lambda_1 | < 1$ they get smaller and smaller, so in either case they will get out of range. To avoid this, a scaling factor is applied after each iteration to standardize $\mathbf{y}^{(m)}$ by making it a unit vector, or by making the numerically largest component equal to 1, or by some other means. Thus

$$\mathbf{y}^{(m+1)} = \frac{1}{k} \mathbf{A}\mathbf{y}^{(m)}$$

where $1/k$ is the scaling factor. When m is large,

$$\mathbf{y}^{(m)} \sim \mathbf{x}^{(1)} \text{ and } k \sim \lambda_1.$$

Thus k converges to the largest latent root and $\mathbf{y}^{(m)}$ converges to the corresponding latent vector. This method is very similar to Bernoulli's method for finding the largest root of a polynomial and most of the methods in the rest of this section are similar to the various remarks and extensions for Bernoulli's method explained in §4.3.

The speed of convergence depends upon the separation of λ_1 and λ_2, i.e. upon the ratio λ_1/λ_2. If $|\lambda_2/\lambda_1| \ll 1$ then convergence will be rapid; on the other hand if $|\lambda_1/\lambda_2| \sim 1$ then convergence will be very slow. There are various ways of accelerating convergence. Thus we could apply Aitken's acceleration process, explained in §4.3, to each component of $\mathbf{y}^{(m)}$. Another method is to shift the origin, i.e. to replace $\mathbf{A}$ by $\mathbf{A} - h\mathbf{I}$ so that the roots become $\lambda_i - h$; it may be possible to choose h so that $\lambda_1 - h$ and $\lambda_2 - h$ remain the numerically largest of the latent roots of $\mathbf{A} - h\mathbf{I}$ and so that

$$(\lambda_2 - h)/(\lambda_1 - h)$$

is numerically much less than 1. A third method is to iterate with $\mathbf{A}^2$ instead of $\mathbf{A}$, for the latent roots of $\mathbf{A}^2$ are

$$\lambda_1^2, \lambda_2^2, \ldots, \lambda_n^2$$

and
$$\lambda_2^2/\lambda_1^2 < \lambda_2/\lambda_1.$$

For example, if $\lambda_2/\lambda_1 \sim 0\cdot9$ then $\lambda_2^2/\lambda_1^2 \sim 0\cdot8$ and convergence will be faster.

Example. To find the largest latent root and corresponding vector of

$$\mathbf{A} = \begin{pmatrix} 1 & 1 & 1 & 1 \\ 1 & 2 & 3 & 4 \\ 1 & 3 & 6 & 10 \\ 1 & 4 & 10 & 20 \end{pmatrix},$$

we have

$$\mathbf{B} = \mathbf{A}^2 = \begin{pmatrix} 4 & 10 & 20 & 35 \\ 10 & 30 & 65 & 119 \\ 20 & 65 & 146 & 273 \\ 35 & 119 & 273 & 517 \end{pmatrix};$$

iterating with **B** we have

y	$\mathbf{y}^{(1)}$	$\mathbf{y}^{(2)}$	$\mathbf{y}^{(3)}$	$\mathbf{y}^{(4)}$	$\mathbf{y}^{(5)}$
1	1·00000	1·00000	1·00000	1·00000	1·00000
1	3·24638	3·34179	3·34247	3·34248	3·34248
1	7·30435	7·60881	7·61100	7·61102	7·61102
1	13·68116	14·35117	14·35117	14·35121	14·35121
$k=$	69	661·39	691·717	691·935	691·93755

Hence the largest root of **B** is 691·93755 and the largest root of **A** is the square root of this number, positive or negative, we check that the right sign is positive and thus

$$\lambda_1 = 26 \cdot 30471, \qquad \mathbf{x}^{(1)} = (1,\ 3 \cdot 34248,\ 7 \cdot 61102,\ 14 \cdot 35121)^{\mathrm{T}}.$$

In this example vectors are scaled so that the first component is 1, this is convenient but not safe as the first component of $\mathbf{x}^{(1)}$ may be zero.

It may happen that the smallest root is required rather than the largest. In that case we iterate with $\mathbf{A}^{-1}$ as the smallest root of **A** is the largest root of $\mathbf{A}^{-1}$; it is possible to do this without computing $\mathbf{A}^{-1}$ explicitly as is done in the example below. More generally, it may happen that the latent root nearest to a certain constant h is required; then we find the numerically smallest root of $\mathbf{A} - h\mathbf{I}$ by iterating with the inverse of $\mathbf{A} - h\mathbf{I}$. This method is called *inverse iteration*.

Example 2. To find the latent root nearest to 2 and the corresponding latent vector for the matrix **A** of example 1, we have:

$$\mathbf{C} = \mathbf{A} - 2\mathbf{I} = \begin{pmatrix} -1 & 1 & 1 & 1 \\ 1 & 0 & 3 & 4 \\ 1 & 3 & 4 & 10 \\ 1 & 4 & 10 & 18 \end{pmatrix} = \mathbf{LU}$$

where, by factorizing, we get

$$\mathbf{L} = \begin{pmatrix} 1 & & & \\ -1 & 1 & & \\ -1 & 4 & 1 & \\ -1 & 5 & 9/11 & 1 \end{pmatrix} \quad \mathbf{U} = \begin{pmatrix} -1 & 1 & 1 & 1 \\ & 1 & 4 & 5 \\ & & -11 & -9 \\ & & & 15/11 \end{pmatrix}$$

Let $\mathbf{D} = \mathbf{C}^{-1}$. Given an arbitrary vector $\mathbf{b}$, to find $\mathbf{Db} = \mathbf{C}^{-1}\mathbf{b}$ we proceed as explained in Remark 4 of the previous section. We solve $\mathbf{Ly} = \mathbf{b}$ for $\mathbf{y}$, then solve $\mathbf{Ux} = \mathbf{y}$ for $\mathbf{x} = \mathbf{C}^{-1}\mathbf{b}$. In the following table we list

$$\mathbf{b} = (b_1, \ b_2, \ b_3, \ b_4)^{\mathrm{T}}$$

$$\mathbf{y} = (y_1, \ y_2, \ y_3, \ y_4)^{\mathrm{T}}$$

$$\mathbf{x} = (x_1, \ x_2, \ x_3, \ x_4)^{\mathrm{T}}$$

Each new $\mathbf{b}$ is obtained from the previous $\mathbf{x}$ by scaling down: $\mathbf{b} = (1/k) \, \mathbf{x}$ so that k tends to the largest latent root of $\mathbf{D}$ and $\mathbf{b}$ to the corresponding vector. As in Example 1 we scale so that b_1 remains unity.

b_1	1	1	1	1
b_2	1	1·303325	1·196116	1·208827
b_3	1	0·837209	0·727917	0·740253
b_4	1	−0·790697	−0·734101	−0·743807
y_1	1	1	1	1
y_2	2	2·302325	2·196116	2·208827
y_3	−6	−7·372091	−7·056547	−7·095055
y_4	−3·090908	−5·270610	−4·941141	−4·982896
x_1	2·866667	5·265111	4·871560	4·921092
x_2	3·733333	6·297669	5·888874	5·940473
x_3	2·4	3·832554	3·606188	3·634741
x_4	−2·266665	−3·865112	−3·623502	−3·654122
$1/k$	0·348837	0·189930	0·205273	0·203207

b_1	1	1	1	1
b_2	1·207146	1·207366	1·207336	1·207341
b_3	0·738605	0·738821	0·738792	0·738797
b_4	−0·742543	−0·742709	−0·742687	−0·742690
y_1	1	1	1	
y_2	2·207146	2·207361	2·207336	
y_3	−7·089979	−7·090643	−7·090552	
y_4	−4·977380	−4·978103	−4·978005	
x_1	4·914544	4·915401	4·915285	
x_2	5·933651	5·934544	5·934423	
x_3	3·630970	3·631463	3·631397	
x_4	−3·650077	−3·650606	−3·650535	
$1/k$	0·203478	0·203442	0·203447	

Thus the largest root of D is $1/0\cdot203447$, hence the smallest root of C is $0\cdot203447$, hence the root of A nearest to 2 is $2\cdot203447$ and the corresponding vector is

$$(1, \ 1\cdot207341, \ 0\cdot738797, \ -0\cdot742690).$$

To find the *subdominant* root λ_2 and corresponding latent vector can be awkward for a general matrix A. But if A is symmetric and we already know the largest latent root λ_1 and corresponding latent vector $x^{(1)}$, we can apply the method of iteration starting with an arbitrary vector y which has no component along $x^{(1)}$, i.e.

$$y = k_2 x^{(2)} + k_3 x^{(3)} + \ldots + k_n x^{(n)}.$$

Then we apply the iteration:

$$y^{(1)} = \frac{1}{k} Ay, \ y^{(2)} = \frac{1}{k} Ay^{(1)}, \ \ldots$$

then $y^{(m)}$ tends to $x^{(2)}$ and k to λ_2. In theory $y^{(1)}, y^{(2)}, \ldots$, should have no component along $x^{(1)}$, but in practice a component along $x^{(1)}$ keeps creeping in and it has to be extracted after each iteration or after every few iterations. The same method can be applied to find λ_3 and $x^{(3)}$; this time we must extract components of $x^{(1)}$

and $\mathbf{x}^{(2)}$. Although in theory we can continue this procedure to find smaller and smaller roots, in practice this becomes too long after, say, 5 or 6 roots; we then have to use some other methods such as shift of origin and inverse iteration as explained above.

Example 3. To find λ_2 and $\mathbf{x}^{(2)}$ for the matrix $\mathbf{A}$ of Example 1 we find λ_2 and $\mathbf{x}^{(2)}$ for the matrix $\mathbf{B} = \mathbf{A}^2$ in order to hasten convergence. In the table below

$$u = \mathbf{y} \, . \, \mathbf{x}^{(1)}/(\mathbf{x}^{(1)} \, . \, \mathbf{x}^{(1)}) = \text{the component of } \mathbf{y} \text{ along } \mathbf{x}^{(1)},$$

$$\mathbf{z} = k'(\mathbf{y} - u\mathbf{x}^{(1)}),$$

so that $\mathbf{z}$ has no component along $\mathbf{x}^{(1)}$, where k' is a scaling factor to make $z_1 = 1$.

$$\mathbf{y}^{(m+1)} = \frac{1}{k} \, \mathbf{A}\mathbf{z}^{(m)}$$

where $1/k$ is a scaling factor to make $y_1^{(m+1)} = 1$

$$
\mathbf{y} =
\begin{matrix}
1 & 1 & 1 & 1 \\
1 & 1\cdot187248 & 1\cdot206234 & 1\cdot207296 \\
1 & 0\cdot712047 & 0\cdot737678 & 0\cdot738738 \\
1 & -0\cdot722365 & -0\cdot741884 & -0\cdot742682
\end{matrix}
$$

$$
k = \qquad\quad 3\cdot390220 \qquad 4\cdot778505 \qquad 4\cdot851550
$$

$$
\mathbf{z} =
\begin{matrix}
1 & 1 & 1 & 1 \\
0\cdot753282 & 1\cdot187084 & 1\cdot206240 & 1\cdot207300 \\
0\cdot303706 & 0\cdot711524 & 0\cdot737695 & 0\cdot738750 \\
-0\cdot406192 & -0\cdot723509 & -0\cdot741850 & -0\cdot742657
\end{matrix}
$$

$$
\mathbf{y} =
\begin{matrix}
1 & 1 \\
1\cdot207324 & 1\cdot207374 \\
0\cdot738751 & 0\cdot738750 \\
-0\cdot742784 & -0\cdot742791
\end{matrix}
$$

$$
k = \quad 4\cdot855005 \qquad 4\cdot855140 \qquad 4\cdot855150
$$

$$
\mathbf{z} =
\begin{matrix}
1 & 1 \\
1\cdot207345 & 1\cdot207338 \\
0\cdot738795 & 0\cdot738796 \\
-0\cdot742690 & -0\cdot742690.
\end{matrix}
$$

Thus

$$\lambda_2(\mathbf{B}) = 4\cdot855150,$$

$$\lambda_2(\mathbf{A}) = \pm\, 4\cdot855150^{\frac{1}{2}} = 2\cdot203444 \text{ (by checking for } \mathbf{A})$$

$$\mathbf{x}^{(2)} = (1,\ 1\cdot207338,\ 0\cdot738796,\ -0\cdot742690)^{\mathrm{T}}$$

which agrees with Example 2 to 5 decimal places.

One of the advantages of the methods of iteration explained in this section is that the latent vectors are found as a by-product of the procedure for finding the latent root, whereas this is not the case in the more advanced methods. The method of this section has much in common with Bernoulli's method for finding the roots of a polynomial explained in §4.3 and it may be thought that a practical way for finding the latent roots of $\mathbf{A}$ is to solve the polynomial equation det $(\mathbf{A} - \lambda\mathbf{I}) = 0$. Unfortunately it is very difficult to find the coefficients of the characteristic polynomial when the order is higher than 3 or 4, and even when it is possible to find the coefficients the numerical calculation is often unstable and gives very inaccurate results. The iterative method fails if $|\lambda_1| = |\lambda_2|$ which happens when $\mathbf{A}$ is real unsymmetric and λ_1, λ_2 are complex conjugate. It is possible to find λ_1, λ_2 by a procedure similar to that explained for the extension of Bernoulli's method in §4.3. When m is sufficiently large

$$\mathbf{y}^{(m)} \sim k_1\lambda_1^m\mathbf{x}^{(1)} + k_2\lambda_2^m\mathbf{x}^{(2)}.$$

Then we can find p, q so that

$$\mathbf{y}^{(m+2)} + p\mathbf{y}^{(m+1)} + q\mathbf{y}^{(m)} \sim 0,$$

and λ_1, λ_2 are the roots of the quadratic equation

$$\lambda^2 + p\lambda + q = 0.$$

EXERCISES

1. Solve the set of equations

$$4x_1 - 4x_2 - 3x_3 + 7x_4 = \;\; 1 \cdot 3$$
$$8x_1 - 3x_2 - 8x_3 + 17x_4 = \;\; 6 \cdot 6$$
$$- 12x_1 + 12x_2 + 16x_3 - 29x_4 = \;\; 2 \cdot 1$$
$$- 8x_1 + 33x_2 - 25x_3 + 36x_4 = 10 \cdot 4$$

by factorizing the matrix of the coefficients in the form LU where **L** is a lower triangular matrix with units on the diagonal and **U** is an upper triangular matrix.   (B.C.)

Ans.

$$\mathbf{L} = \begin{pmatrix} 1 & 0 & 0 & 0 \\ 2 & 1 & 0 & 0 \\ -3 & 0 & 1 & 0 \\ -2 & 5 & -3 & 1 \end{pmatrix}, \; \mathbf{U} = \begin{pmatrix} 4 & -4 & -3 & 7 \\ 0 & 5 & -2 & 3 \\ 0 & 0 & 7 & -8 \\ 0 & 0 & 0 & 11 \end{pmatrix}, \; \mathbf{x} = \begin{pmatrix} 1 \cdot 075 \\ 1 \\ 2 \\ 1 \end{pmatrix}$$

2. If

$$\mathbf{X} = \begin{pmatrix} 2 & -2 & 1 \\ 1 & 2 & 2 \\ 2 & 1 & -2 \end{pmatrix}, \quad \mathbf{A} = \begin{pmatrix} 25/9 & 2/9 & -8/9 \\ 2/9 & 34/9 & -10/9 \\ -8/9 & -10/9 & 31/9 \end{pmatrix}$$

calculate $\mathbf{X}^{-1}$ and prove that $\mathbf{X}^{-1}\mathbf{A}\mathbf{X} =$ a diagonal matrix of the form

$$\begin{pmatrix} \lambda_1 & 0 & 0 \\ 0 & \lambda_2 & 0 \\ 0 & 0 & \lambda_3 \end{pmatrix}.$$

Hence show that

$$(\mathbf{A} - \lambda_1\mathbf{I})(\mathbf{A} - \lambda_2\mathbf{I})(\mathbf{A} - \lambda_3\mathbf{I}) = 0. \qquad \text{(B.C.)}$$

Hint. If $\mathbf{D} = \mathbf{X}^{-1}\mathbf{A}\mathbf{X}$ is the diagonal matrix, then $\mathbf{A} = \mathbf{X}\mathbf{D}\mathbf{X}^{-1}$ and

$$(\mathbf{A} - \lambda\mathbf{I}) = \mathbf{X}(\mathbf{D} - \lambda\mathbf{I})\mathbf{X}^{-1}$$

3. Solve the following set of equations by elimination with interchanges.

$$1595x_1 - 3465x_2 + 2475x_3 - 585x_4 = 2000$$
$$- 585x_1 + 2475x_2 - 3465x_3 + 1595x_4 = -2000$$
$$2475x_1 - 8505x_2 + 9515x_3 - 3465x_4 = 6000$$
$$- 3465x_1 + 9515x_2 - 8505x_3 + 2475x_4 = -6000$$

Ans. The process of elimination with interchanges produces

$$L = \begin{pmatrix} 1 & & & \\ 0 \cdot 1688 & 1 & & \\ -0 \cdot 7143 & -0 \cdot 5084 & 1 & \\ -0 \cdot 4603 & -0 \cdot 5355 & -0 \cdot 6972 & 1 \end{pmatrix},$$

$$U = \begin{pmatrix} -3465 & 9515 & -8505 & 2475 \\ & -1708 \cdot 57 & 3440 & -1697 \cdot 14 \\ & & 402 \cdot 08 & -354 \cdot 51 \\ & & & 67 \cdot 22 \end{pmatrix},$$

$$\mathbf{x} = (0 \cdot 1, \ -0 \cdot 3, \ 0 \cdot 3, \ -0 \cdot 1)^T$$

4. Find the latent roots and vectors of the following matrix:

$$\begin{pmatrix} 1 & 1 & 1 & 1 \\ 1 & 2 & 2 & 2 \\ 1 & 2 & 3 & 3 \\ 1 & 2 & 3 & 4 \end{pmatrix}.$$

(U.L.)

Ans.

$$\lambda = \quad 8 \cdot 290859 \quad 1 \quad\quad 0 \cdot 426022 \quad 0 \cdot 283119$$

$$\mathbf{x} = \begin{pmatrix} 0 \cdot 228013 & 0 \cdot 577350 & 0 \cdot 656538 & -0 \cdot 428525 \\ 0 \cdot 428525 & 0 \cdot 577350 & -0 \cdot 228013 & 0 \cdot 656538 \\ 0 \cdot 577350 & 0 & -0 \cdot 577350 & -0 \cdot 577350 \\ 0 \cdot 656538 & -0 \cdot 577350 & 0 \cdot 428528 & 0 \cdot 228013 \end{pmatrix}$$

5. Find the latent roots and vectors of the following matrix:

$$\begin{pmatrix} 2 & -1 & 0 & 0 \\ -1 & 2 & -1 & 0 \\ 0 & -1 & 2 & -1 \\ 0 & 0 & -1 & 2 \end{pmatrix}.$$

Ans.

$$\lambda = \quad 3\cdot618034 \quad 2\cdot618034 \quad 1\cdot381966 \quad 0\cdot381966$$

$$\mathbf{x} = \begin{pmatrix} 0\cdot371748 & 0\cdot601501 & -0\cdot601501 & 0\cdot371748 \\ -0\cdot601501 & -0\cdot371748 & -0\cdot371748 & 0\cdot601501 \\ 0\cdot601501 & -0\cdot371748 & 0\cdot371748 & 0\cdot601501 \\ -0\cdot371748 & 0\cdot601501 & 0\cdot601501 & 0\cdot371748 \end{pmatrix}$$

CHAPTER 7

Relaxation Methods

WE EXPLAIN in this chapter iterative methods which can be applied in many contexts. For purposes of illustration we choose three problems: the solution of a system of linear algebraic equations, the solution of second order ordinary differential equations with two-point boundary conditions, and the solution of Laplace's or Poisson's elliptic partial differential equation. The same concepts and terms are used for these apparently different problems. In each case we give two approaches: one suitable for hand or desk computations and one for digital computers. Although the digital computer methods are obviously more important, the beginner would be well advised to try some hand computation methods in order to acquire some experience and insight into relaxation problems and because these techniques are often the starting point of the more important methods.

7.1 Solution of Linear Equations

For illustration we take three equations in three unknowns

$$\left.\begin{array}{l} a_{11}x_1 + a_{12}x_2 + a_{13}x_3 = b_1, \\ a_{21}x_1 + a_{22}x_2 + a_{23}x_3 = b_2, \\ a_{31}x_1 + a_{32}x_2 + a_{33}x_3 = b_3, \end{array}\right\} \qquad (7.1.1)$$

but it should be understood that the methods explained here are usually applied when the number of equations n is large. We define the *residuals* r_1, r_2, r_3 by the relation

$$r_1 = a_{11}x_1 + a_{12}x_2 + a_{13}x_3 - b_1$$

with similar expressions for r_2 and r_3. The object is to obtain better and better approximations to (x_1, x_2, x_3) so as to "liquidate" the residuals, i.e. to make the residuals r_1, r_2, r_3 smaller and smaller and eventually negligible. With this object in view we construct an "operations table" showing the effect on r_1, r_2, r_3 when x_1 is given an increment of 1, keeping the other unknowns constant, calling this operator R_1, and similarly for x_2 and x_3. Note that the entries in the following operations table represent *changes* in the x's and the resulting *changes* in the r's.

	x_1	x_2	x_3	r_1	r_2	r_3	
R_1	1	0	0	a_{11}	a_{21}	a_{31}	
R_2	0	1	0	a_{12}	a_{22}	a_{32}	(7.1.2)
R_3	0	0	1	a_{13}	a_{23}	a_{33}	

Thus the *operator* R_2 increases x_2 by 1, keeping x_1 and x_3 constant. This has the effect of increasing r_1, r_2, r_3 by a_{12}, a_{22}, a_{32} respectively. If we apply the operator $9R_2$ this increases x_2 by 9 units and r_1, r_2, r_3 by $9a_{12}$, $9a_{22}$, $9a_{32}$ respectively. We observe that the operations table (7.1.2) consists of the unit matrix $\mathbf{I}$ and $\mathbf{A}^T$, the transpose of the matrix $\mathbf{A}$ of the coefficients in (7.1.1).

Example. To solve the equations:

$$10x_1 - 2x_2 + x_3 = 12,$$
$$x_1 + 9x_2 - x_3 = 10,$$
$$2x_1 - x_2 + 11x_3 = 20,$$

we have the operation table:

	x_1	x_2	x_3	r_1	r_2	r_3
R_1	1	0	0	10	1	2
R_2	0	1	0	-2	9	-1
R_3	0	0	1	1	-1	11

The actual relaxation table is given below with explanatory notes.

x_1	x_2	x_3	r_1	r_2	r_3	
0	0	0	-12	-10	-20	(1)
0	0	2	-10	-12	2	(2)
0	1	0	-12	-3	1	(3)
1	0	0	-2	-2	3	(4)
1	1	2	-2	-2	3	(5)
10	10	20	-20	-20	30	(6)
0	0	-3	-23	-17	-3	(7)
2	0	0	-3	-15	1	(8)
0	2	0	-7	3	-1	(9)
1	0	0	3	4	1	(10)
13	12	17	3	4	1	(11)

(1) This line is obtained by taking $x_1 = x_2 = x_3 = 0$ and calculating r_1, r_2, r_3. Any other convenient approximation could be used instead.

(2) Observing that the numerically largest residual is r_3 we try to liquidate it by applying a *simple* (i.e. one figure) multiple of the operator R_3. In this case we apply $2R_3$; from the operator table we deduce that this adds 2, -2, 22 to r_1, r_2, r_3 respectively to obtain the values of r_1, r_2, r_3 in line (2).

(3–4) These lines are obtained similarly by applying $1R_2$ and $1R_1$.

(5) This line is obtained by adding the increments of x_1, x_2, x_3 in lines (2), (3), (4) to the initial approximations to x_1, x_2, x_3 in line (1). Line (5) is a "stock-taking" stage where the residuals r_1, r_2, r_3 are obtained by substituting $x_1 = 1$, $x_2 = 1$, $x_3 = 2$ in the given equations and we check that we obtain the same r_1, r_2, r_3 as in line (4).

(6) We now reach a stage where in order to proceed further we must introduce decimal fractions; to avoid this we multiply line (5) by 10 to obtain line (6) where the units now represent the first decimal place. This line is normally combined with line (5).

(7)–(10) These lines are obtained by applying $-3R_3$, $2R_1$, $2R_2$, $1R_1$, respectively.

(11) This is a stock-taking line where the current values of x_1, x_2, x_3 are obtained by adding the increments from lines (7)–(10) to the starting values in line (6). Thus line (11) gives $x_1 = 1\cdot3$, $x_2 = 1\cdot2$, $x_3 = 1\cdot7$ and we check by substitution in the given equations that $r_1 = 0\cdot3$, $r_2 = 0\cdot4$, $r_3 = 0\cdot1$. If we obtain and verify different values for $10r_1$, $10r_2$, $10r_3$ than the ones given in line (10) we need not repeat the calculations of earlier lines; we simply continue with the corrected residuals. To obtain two decimal accuracy we multiply line (11) by 10 and continue.

Thus it is possible to obtain results correct to 6 or 7 decimal places, yet the calculations involve 1- or at most 2-digit numbers only, except at stock-taking. Operations need not be simple changes in one variable at a time. Thus if we have the following operations table:

	x_1	x_2	x_3	r_1	r_2	r_3
R_1	1	0	0	10	9	-11
R_2	0	1	0	5	4	3
R_3	0	0	1	7	7	1
we define						
R_4	1	-2	0	0	1	-17
R_5	1	5	-4	7	1	0
R_6	-1	-5	5	0	6	1

where

$$R_4 = R_1 - 2R_2$$
$$R_5 = R_1 + 5R_2 - 4R_3$$
$$R_6 = -R_1 - 5R_2 + 5R_3$$

The operators R_4, R_5, R_6 are more suitable for liquidating the residuals one at a time. Thus R_4 can be used to liquidate r_3 with-

7

out altering r_1 and with only a small change in r_2. This method of combining the given basic operators to achieve a desired effect is called the method of block relaxation or group operation.

On a digital computer this work would be carried out differently as it no longer saves time to restrict the arithmetic to small integers. One approach is to determine the numerically largest of the residuals $r_1, r_2, \ldots, r_n$, where n is the number of equations. If, say, r_2 is the largest then a new value of x_2 is obtained by solving the second equation of (7.1.1) for x_2 so that r_2 becomes zero; then calculate or readjust the other residuals r_1 and r_3 for the new approximate solution (x_1, x_2, x_3). This method is, however, not very popular because the process of scanning a large set of numbers $r_1, r_2, \ldots, r_n$, to find the numerically largest is relatively time-consuming on a digital computer. Instead the residuals $r_1, r_2, \ldots, r_n$, are systematically liquidated in that order. Let $n = 3$ and let

$$\mathbf{x}^{(m)} = (x_1^{(m)}, \ x_2^{(m)}, \ x_3^{(m)})$$

be the solution after the mth iteration. Then the procedure may be described in matrix form as follows

$$\left. \begin{aligned} a_{11}x_1^{(m+1)} + a_{12}x_2^m + a_{13}x_3^m &= b_1, \\ a_{21}x_1^{(m+1)} + a_{22}x_2^{(m+1)} + a_{23}x_3^m &= b_2, \\ a_{31}x_3^{(m+1)} + a_{32}x_2^{(m+1)} + a_{33}x_3^{(m+1)} &= b_3. \end{aligned} \right\} \qquad (7.1.3)$$

This system is solved by forward substitution

$$\left. \begin{aligned} x_1^{(m+1)} &= (b_1 - a_{12}x_2^{(m)} - a_{13}x_3^{(m)})/a_{11}, \\ x_2^{(m+1)} &= (b_2 - a_{21}x_1^{(m+1)} - a_{23}x_3^{(m)})/a_{22}, \\ x_3^{(m+1)} &= (b_3 - a_{31}x_1^{(m+1)} - a_{32}x_2^{(m+1)})/a_{33}. \end{aligned} \right\} \qquad (7.1.4)$$

Let $\mathbf{A}$ be the matrix of the coefficients and $\mathbf{L}$, $\mathbf{U}$ defined as follows

$$\mathbf{L} = \begin{pmatrix} a_{11} & 0 & 0 \\ a_{21} & a_{22} & 0 \\ a_{31} & a_{32} & a_{33} \end{pmatrix}, \qquad \mathbf{U} = \begin{pmatrix} 0 & a_{12} & a_{13} \\ 0 & 0 & a_{23} \\ 0 & 0 & 0 \end{pmatrix}$$

so that $$\mathbf{A} = \mathbf{L} + \mathbf{U}.$$

Then (7.1.3) is equivalent to the equation

$$\mathbf{Lx}^{(m+1)} + \mathbf{Ux}^{(m)} = \mathbf{b}$$

and (7.1.4) is equivalent to the equation

$$\mathbf{x}^{(m+1)} = \mathbf{L}^{-1}(\mathbf{b} - \mathbf{Ux}^{(m)}).$$

This is the *Gauss–Seidel* method. It is essentially similar to a relaxation method with the residuals liquidated in turn instead of liquidating the largest, although in fact the residuals are not explicitly calculated.

Example. To solve the equations of the previous example by this method:

$$10x_1 - 2x_2 + x_3 = 12,$$

$$x_1 + 9x_2 - x_3 = 10,$$

$$2x_1 - x_2 + 11x_3 = 20,$$

we have

$$x_1^{(m+1)} = 1 \cdot 2000 + 0 \cdot 2000x_2^{(m)} - 0 \cdot 1000x_3^{(m)},$$

$$x_2^{(m+1)} = 1 \cdot 1111 - 0 \cdot 1111x_1^{(m+1)} + 0 \cdot 1111x_3^{(m)},$$

$$x_3^{(m+1)} = 1 \cdot 8182 - 0 \cdot 1818x_1^{(m+1)} + 0 \cdot 0909x_2^{(m+1)}.$$

m	1	2	3	4	5	6
$x_1^{(m)}$	0	$1 \cdot 2000$	$1 \cdot 2267$	$1 \cdot 2624$	$1 \cdot 2625$	$1 \cdot 2624$
$x_2^{(m)}$	0	$0 \cdot 9778$	$1 \cdot 1624$	$1 \cdot 1598$	$1 \cdot 1591$	$1 \cdot 1591$
$x_3^{(m)}$	0	$1 \cdot 6889$	$1 \cdot 7008$	$1 \cdot 6941$	$1 \cdot 6940$	$1 \cdot 6941$

Thus the required solution is $x_1 = 1 \cdot 2624$, $x_2 = 1 \cdot 1591$, $x_3 = 1 \cdot 6941$ correct to four decimal places.

In the relaxation methods explained above, each equation of the set is associated with one of the variables for which it is solved. In physical applications this association is usually obvious. With this arrangement the diagonal elements of the matrix **A** of the coefficients must *dominate* the other coefficients in the corresponding row, if the relaxation methods are to succeed. To be more precise, for $n = 3$, we must have

$$| a_{11} | \geqslant | a_{12} | + | a_{13} |$$
$$| a_{22} | \geqslant | a_{21} | + | a_{23} |$$
$$| a_{33} | \geqslant | a_{31} | + | a_{32} |$$

with strict inequality for at least one row; the proof of this condition for the convergence of relaxation methods is beyond the scope of this book. The speed of convergence depends upon how dominating the diagonal elements are; if the diagonal elements are not sufficiently large then the relaxation methods will not converge at all; if the equality, rather than the inequality, holds in most of the above relations, then convergence is very slow. Unfortunately this is frequently the case in applications and many advanced iterative techniques have been developed to deal with this problem; one such method is explained in §7.3. In general a set of n linear equations in n unknowns is solved by relaxation methods rather than by the direct methods of elimination of §6.1 when n is large and the diagonal elements dominate. Relaxation methods are particularly suitable when the matrix of the coefficient is *sparse*, i.e. many of the coefficients are zero.

7.2 Second Order Ordinary Differential Equations with Two-point Boundary Conditions

The problem is to solve the differential equation

$$y'' = f(x, y, y')$$

in the range $(a \leqslant x \leqslant b)$ given the values of y at $x = a$ and $x = b$.

To illustrate the main ideas we first study a special case by taking the differential equation in the special form

$$y'' = f(x),$$

although this could be solved by integration; later on we consider more general equations.

We divide the range into $n + 1$ intervals of length h so that there are n internal points in the range:

$$a + h, \; a + 2h, \; \ldots, \; a + nh.$$

Let x_0 be any of these internal points and $x_{-1} = x_0 - h$, $x_1 = x_0 + h$, $\ldots$. Let the solutions at x_{-1}, x_0, x_1 be denoted by y_{-1}, y_0, y_1 respectively. By (3.8.3) we have

$$h^2 y'' = \delta^2 y - \tfrac{1}{12}\delta^4 y + \; \ldots$$

As a first approximation we take

$$h^2 y_0'' = \delta^2 y_0 = y_{-1} - 2y_0 + y_1.$$

Thus the differential equation $y'' = f$ becomes, when applied at x_0,

$$y_{-1} - 2y_0 + y_1 = h^2 f_0. \qquad (7.2.1)$$

The residual $r(x_0)$ is defined by the equation

$$r(x_0) = y_{-1} - 2y_0 + y_1 - h^2 f_0. \qquad (7.2.2)$$

We examine the effect on $r(x_0)$, $r(x_{-1})$, $r(x_1)$ when we give y_0 an increment of 1:

$$r(x_0) \text{ decreases by } 2$$

as is obvious from (7.2.2).

$$r(x_{-1}) \text{ increases by } 1$$

because y_0 relative to x_0 becomes y_1 relative to x_{-1}. Similarly $r(x_1)$ increases by 1. We call this effect the *relaxation operator* R_0 and represent it by the diagram:

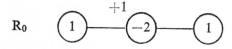

If, however, the point x_0 is the point $a + h$ next to the left boundary then $x_{-1} = a$ and $y_{-1} = y(a)$ is fixed so that there is no residual at that point; then the operator R_0 becomes the operator R_a defined in the diagram. Similarly we define R_b for the point $b - h$ next to the right boundary point.

To solve the differential equation we proceed as follows: given $y(a)$ and $y(b)$ we divide the range (a, b) into $n + 1$ intervals and guess the values of y at the n internal points. By (7.2.2) we calculate the residual r at every internal point.

We then pick out the largest residual and apply a suitable multiple of the corresponding relaxation operation R_0, R_a or R_b to liquidate that residual and repeat this process, readjusting the values of y so as to make the residuals as small as possible. Much of the procedure and remarks of the previous section apply here also. Further, besides attempting to make the residuals as small as possible individually, we also try to keep sums of residuals, over any set of consecutive points, reasonably small, i.e. we must avoid residuals of similar signs as they give a "bow-window" or "pillow" effect of errors in the required solution y and although the residuals may be small the error in the solution will be large, particularly towards the middle of the range. This last idea is called "over-relaxation", because it usually involves larger increments in y than is necessary merely to liquidate the residual at a particular point.

Example. To solve the differential equation

$$y'' = \tfrac{1}{4}\pi^2 \sin (\tfrac{1}{2}\pi x), \qquad y(0) = 0 \cdot 5, \qquad y(1) = -0 \cdot 5.$$

Here $a = 0$ and $b = 1$; we take $h = 0 \cdot 25$ so that there are three internal points. We guess $y(0 \cdot 25) = 0 \cdot 25$, $y(0 \cdot 5) = 0$, $y(0 \cdot 75) = -0 \cdot 25$ by simple linear interpolation. Equation (7.2.2) becomes

$$r(x_0) = y_{-1} - 2y_0 + y_1 - \tfrac{1}{64}\pi^2 \sin (\tfrac{1}{2}\pi x_0)$$

which gives

$$r(0 \cdot 25) = -0 \cdot 6, \quad r(0 \cdot 5) = -0 \cdot 11, \quad r(0 \cdot 75) = -0 \cdot 14.$$

The computation is illustrated in Fig. 3. We carry out the arithmetic to two decimal places and to avoid decimal fractions we

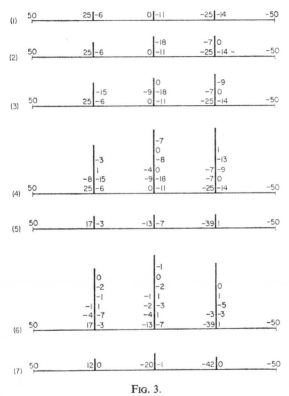

Fig. 3.

multiply the y's and the residuals by 100 to get line (1). There is a vertical line through each internal point; on its left we record the value of y at that point and any increments we may give it, on its right we record the residual. At the end points we record the given boundary values. At each stage we select the point with the largest residual and apply a suitable multiple of the appropriate relaxation operator R_0 or R_a or R_b. Thus applying $-7R_b$ at $0 \cdot 75$ we get line (2); applying $-9R_0$ at $0 \cdot 5$, line (2) becomes line (3). On the computation sheet we have only one diagram which starts as in line (1), and is built up to form (2), (3), ... , and so on. At each internal point x_0 there is a vertical line with a growing stack of numbers on each side. The current residual at x_0 is the *actual top number* in the right stack, whereas the current value of y_0 is the *sum of the numbers* in the left stack. When one of the stacks becomes too high we start a fresh line with the latest y's and residuals; at this stage we check the residuals and correct if necessary. Thus applying $-7R_b$ at $0 \cdot 75$, $-9R_0$ at $0 \cdot 5$, $-8R_a$ at $0 \cdot 25$, $-4R_0$ at $0 \cdot 5$, $-7R_b$ at $0 \cdot 75$, line (1) becomes line (4). Starting afresh we get line (5). Applying $-4R_0$, $-4R_a$, $-2R_0$, $-3R_b$, $-1R_0$, $-1R_a$, line (5) becomes line (6). In line (7) we record the latest y's and residuals, and check. It is emphasized here that only lines (4) and (6) need appear on the computation sheet; the other lines represent intermediate stages shown for illustration. Thus the solution is

$$y(0 \cdot 25) = 0 \cdot 12, \quad y(0 \cdot 5) = -0 \cdot 20, \quad y(0 \cdot 75) = -0 \cdot 42.$$

The exact solution is $y = \frac{1}{2} - \sin\left(\frac{1}{2}\pi x\right)$ which gives

$$y(0 \cdot 25) = 0 \cdot 12, \quad y(0 \cdot 5) = -0 \cdot 21, \quad y(0 \cdot 75) = -0 \cdot 42.$$

The results obtained by relaxation are remarkably accurate for such a coarse interval $h = 0 \cdot 25$. In general the solution is not so easy; to improve accuracy we can add more digits and recompute residuals with extra figures; we can introduce more points by reducing the interval; to improve accuracy we also have to take into account more terms, $-\frac{1}{12}\delta^4 + \ldots$, applied as correction terms (see Remark 2 below). To speed up the relaxation process we can apply block relaxation, for example we can apply a relaxation operator involving equal changes in 2 or 3 consecutive points. In the remarks below we consider differential equations less trivial than the above example and methods suitable for digital computers.

Remark 1. Let **y** be the vector $(y_1, y_2, \ldots, y_n)^T$ where y_i is the solution at the internal point $x = a + ih$. This is a slight change of notation. When we apply equation (7.2.1) at the n internal points we get a system of linear equations of the form

$$\mathbf{Ay} = \mathbf{b}. \qquad (7.2.3)$$

Here **A** is a *co-diagonal* matrix where all the diagonal elements are -2, the elements immediately preceding and succeeding the diagonal elements are 1, and all the other elements are zero. For example, if $n = 5$ then

$$\mathbf{A} = \begin{pmatrix} -2 & 1 & 0 & 0 & 0 \\ 1 & -2 & 1 & 0 & 0 \\ 0 & 1 & -2 & 1 & 0 \\ 0 & 0 & 1 & -2 & 1 \\ 0 & 0 & 0 & 1 & -2 \end{pmatrix}$$

This type of matrix is very frequent in application, it is sometimes called a *band matrix* because all the elements are zero except inside a band along the diagonal. This simple band matrix has width 3. For, say, $n = 100$ (which is the sort of size which may be needed in applications) **A** has 10,000 elements, all zero except 298 of them.

Returning to (7.2.3), the right-hand vector

$$\mathbf{b} = (b_1, b_2, \ldots, b_n)^T$$

is obtained from the formula

$$b_i = h^2 f(a + ih) = h^2 f_i, \qquad (i = 2, 3, \ldots, n - 1)$$
$$b_1 = h^2 f_1 - y(a),$$
$$b_n = h^2 f_n - y(b).$$

The equation (7.2.3) can be solved by the relaxation method of the previous section; the method of this section is in fact equivalent to doing this for the particular matrix $\mathbf{A}$ we have here. Alternatively we can solve (7.2.3) by the method of elimination of §6.2.

Remark 2. So far in this section the differential equation $y'' = f(x)$ has been solved by using the approximation $h^2 y'' = \delta^2 y$. To improve on the solution we should take the next term in the approximation,

$$h^2 y'' = \delta^2 y - \tfrac{1}{12}\delta^4,$$

from (3.8.3). Now the relation $\delta^2 y = h^2 f$ in matrix form is equivalent to $\mathbf{Ay} = \mathbf{b}$ as in (7.2.3), where $\mathbf{Ay}$ is essentially $\delta^2 y$ and $\mathbf{b}$ is essentially $h^2 f$. Thus the general row of the matrix $\mathbf{A}$ is

$$(0,\ 0,\ \ldots,\ 0,\ 1,\ -2,\ 1,\ 0,\ \ldots,\ 0).$$

This corresponds to the coefficients in the formula

$$\delta^2 y_0 = y_{-1} - 2y_0 + y_1.$$

In a similar way $\delta^4 y$ is equivalent to $\mathbf{Cy}$ where $\mathbf{C}$ is a band matrix of width 5 whose general row is of the form

$$(0,\ 0,\ \ldots,\ 0,\ 1,\ -4,\ 6,\ -4,\ 1,\ 0,\ \ldots,\ 0)$$

corresponding to the coefficients in the formula

$$\delta^4 y_0 = y_{-2} - 4y_{-1} + 6y_0 - 4y_1 + y_2.$$

Thus a more accurate approximation to the differential equation $y'' = f(x)$ is the equation

$$\delta^2 y = \tfrac{1}{12}\delta^4 y = h^2 f,$$

which becomes in matrix form

$$\mathbf{Ay} - \tfrac{1}{12}\mathbf{Cy} = \mathbf{b}.$$

Here the vector $\mathbf{b} = (b_1,\ b_2,\ \ldots,\ b_n)^{\mathrm{T}}$ is essentially $h^2 f(x)$ evaluated at the n internal points, but the first two components b_1, b_2 and the last two components b_{n-1}, b_n need special attention which will not be described here. The improved solution is obtained by solving the linear equation $\mathbf{By} = \mathbf{b}$ where $\mathbf{B} = \mathbf{A} - \frac{1}{12}\mathbf{C}$. However, when n is large $\mathbf{By} = \mathbf{b}$ is much more awkward to solve than $\mathbf{Ay} = \mathbf{b}$.

Instead we can use an iterative method which is essentially the same as the Fox–Goodwin method explained in §5.4. We consider successive approximations $\mathbf{y}^{(1)}$, $\mathbf{y}^{(2)}$, $\ldots$, $\mathbf{y}^{(m)}$, $\mathbf{y}^{(m+1)}$, $\ldots$. We emphasize that $\mathbf{y}^{(m)}$ is a vector whose components are to be the solution of the differential equation at the n internal points. We obtain $\mathbf{y}^{(1)}$ by solving the equation $\mathbf{Ay} = \mathbf{b}$. In general $\mathbf{y}^{(m+1)}$ is obtained from $\mathbf{y}^{(m)}$ by solving for $\mathbf{y}$ the equation

$$\mathbf{Ay} = \mathbf{b} + \tfrac{1}{12}\mathbf{Cy}^{(m)}.$$

The iteration is stopped when $\mathbf{y}^{(m+1)}$ and $\mathbf{y}^{(m)}$ agree to the desired accuracy or when $\mathbf{Ay} - \frac{1}{12}\mathbf{Cy} - \mathbf{b}$ is sufficiently small. Thus we are solving the same matrix $\mathbf{A}$ but for different right-hand sides, and $\mathbf{A}$ is factorized once and for all in the form $\mathbf{LU}$ (see §6.2 and Example 2 of §6.3). Because $\mathbf{A}$ is a band matrix, $\mathbf{L}$ and $\mathbf{U}$ will also be band matrices and in fact, if the factorization is done without interchanges, $\mathbf{L}$ and $\mathbf{U}$ together form the same band structure and take up the same storage space as $\mathbf{A}$.

Remark 3. We now show how to extend the methods of this section to differential equations more complicated than $y'' = f(x)$. For illustration we consider the differential equation

$$y'' + g(x)\,y = f(x).$$

Using the approximation $h^2 y'' = \delta^2 y$, the equation becomes

$$\delta^2 y + h^2 g y = h^2 g,$$

or
$$y_{-1} - (2 - h^2 g)\,y_0 + y_1 = h^2 f.$$

Thus all the methods explained above could be applied in exactly the same manner except that -2 becomes $-(2 - h^2 g)$ where g stands for $g(x)$ evaluated at the point

$$x_0 = a + h,\ a + 2h,\ \ldots,\ a + nh.$$

Thus the relaxation operator

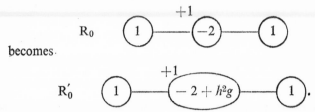

R_0

becomes

R_0'

When using hand computation we can simplify the arithmetic considerably as follows. Usually h^2g is small as compared with -2; thus we can replace R_0' by R_0 when liquidating the residuals and take the term h^2g into account only when computing the residual at the beginning and when checking at a stock-taking stage.

Similarly the matrix methods of the previous remarks can still be applied except that now the diagonal terms are no longer all equal to -2 but to $-2 + h^2g$ where $g = g(x)$ varies from one point to another.

Example. We illustrate the various ideas of this section by applying them to a type of two-point boundary problem called the *eigenvalue problem*. Consider for example the differential equation

$$y'' + \lambda(1 + x^2)\, y = 0, \qquad y(-1) = 0, \qquad y(1) = 0.$$

We seek a solution in the range $(-1, 1)$; the situation is different from the previous example because here the boundary values are zero. This differential equation has a non-zero solution only for certain values of λ called the *eigenvalues* and the corresponding solutions are called *eigenfunctions*. This can be seen by considering the finite difference approximation

$$\delta^2 y + h^2\lambda(1 + x^2)\, y = 0.$$

We take

$$h = 0 \cdot 5, \qquad x_1 = -0 \cdot 5, \qquad x_2 = 0, \qquad x_3 = 0 \cdot 5$$

and let the corresponding values of y be (y_1, y_2, y_3). Then we can express the finite difference approximation in the following matrix form

$$\mathbf{A'y} + h^2\lambda\mathbf{B'y} = 0,$$

where

$$\mathbf{A'} = \begin{pmatrix} -2 & 1 & 0 \\ 1 & -2 & 1 \\ 0 & 1 & -2 \end{pmatrix}, \quad \mathbf{y} = \begin{pmatrix} y_1 \\ y_2 \\ y_3 \end{pmatrix}, \quad \mathbf{B'} = \begin{pmatrix} 1\cdot25 & 0 & 0 \\ 0 & 1 & 0 \\ 0 & 0 & 1\cdot25 \end{pmatrix}.$$

To simplify the arithmetic we replace this problem by the following:

$$\mathbf{Ay} = k\mathbf{By}$$

where

$$\mathbf{A} = -\mathbf{A'}, \quad k = \frac{\lambda}{16}, \quad \mathbf{B} = 4\mathbf{B'} = \begin{pmatrix} 5 & 0 & 0 \\ 0 & 4 & 0 \\ 0 & 0 & 5 \end{pmatrix}.$$

This problem is very similar to the problem of latent roots and vectors of §6.3; it is a little different because **B** is not quite a unit matrix, but most of the results of §6.3 still hold here.

This type of problem frequently arise in the study of vibrations of mechanical systems. We seek the smallest eigenvalue and the corresponding eigenfunction; this represents the smallest natural frequency and the corresponding mode.

For a given approximation **y** we can get a good approximation to k from a simple extension of the Rayleigh–Ritz formula of §6.3:

$$k \sim \frac{\mathbf{y}^T\mathbf{Ay}}{\mathbf{y}^T\mathbf{By}}$$

With this value of k and the given approximation to y we compute

$$\mathbf{b} = k\mathbf{By}$$

and seek to solve the equation

$$\mathbf{Ay} = \mathbf{b}$$

by minimizing the residual

$$R = Ay - b$$

by applying relaxation operators of the type R_a, R_0, R_b. We denote the vector of the resulting increments by δy. We then compute the improved approximation

$$y' = y + \delta y$$

and the reduced residual

$$R' = Ay' - b.$$

We can now calculate again an improved approximation to k using y' and continue iteratively. This is done in the computations below and followed by some explanation of further points in the computation procedure.

	First approximation	Second approximation	Third approximation
y	100	77	74
	100	100	100
	100	77	74
Ay	100	54	48
	0	46	52
	100	54	48
y^TAy	20 000	12 916	12 304
By	500	385	370
	400	400	400
	500	385	370
y^TBy	140 000	99 290	94 760
k	0·14	0·130	0·130
b	70	50	48
	56	52	52
	70	50	48
R	30	4	0
	−56	−6	0
	30	4	0
δy	−23	−3	
	0	0	
	−23	−3	
y'	77	74	
	100	100	
	77	74	
R'	−16	−2	
	−10	0	
	−16	−2	

We now remark on some points of detail in the above computation. We make considerable use of the symmetry of the solution with respect to $x = 0$. As an eigenvector is determined up to a constant multiplier, we fix y_2 to a constant value $= 100$. We choose $\delta\mathbf{y}$ so that the new residual $\mathbf{R}'$ is as small as possible; we also try to arrange as far as possible that $\mathbf{R}'$ is proportional to the right-hand side $\mathbf{b}$; this will speed up the convergence of k. In this example we aim at two-figure accuracy; to simplify the arithmetic and avoid decimal places we deal, whenever possible, with integers only, multiplying y by a suitable power of 10. The final answer is that the smallest eigenvalue is

$$\lambda = 16k = 2{\cdot}08.$$

The corresponding eigenfunction is:

x	-1	$-0{\cdot}5$	0	$0{\cdot}5$	1
y	0	74	100	74	0

We cannot get better accuracy by introducing more figures unless we also increase the number of intervals and possibly use more terms in the finite difference approximation to y'', using the Fox–Goodwin method to deal with the extra terms as correction terms.

On a digital computer this problem could be solved by a suitable adaptation of the method of inverse iteration explained in Example 2 of §6.3; the method of the present example is equivalent to a way of carrying out inverse iteration using relaxation methods and hand computation only.

7.3 Solution of Partial Differential Equations of the Poisson Type

We extend the methods of the previous section to Poisson's partial differential equation in two dimensions:

$$\frac{\partial^2 u}{\partial x^2} + \frac{\partial^2 u}{\partial y^2} = f(x, y) \qquad (7.3.1)$$

Laplace equation is a special case of this when f is replaced by 0. The methods explained here apply to a wider class of partial differential equations of the elliptic type. The problem we consider is to find u inside an area in the (x, y) plane, given its value on the boundary of this area. For simplicity we take the area to be a square and divide it by horizontal and vertical lines into a square net of mesh size h, so that if the origin is at one corner, the co-ordinates of a *nodal-* or *mesh-point* are (ih, jh) or (i, j) for short. If each side is divided into $n + 1$ intervals, there are n^2 internal points and $4n + 4$ boundary points. For the boundary points $i = 0$ or $n + 1$ or $j = 0$ or $n + 1$. The problem then is to determine u at each of the internal nodes, given u at the boundary nodes.

We consider any internal point (x_0, y_0). For brevity we denote

$$(x_0, y_0), (x_0 + h, y_0), (x_0, y_0 + h), (x_0 - h, y_0), (x_0, y_0 - h)$$

by   0,   1,   2,   3,   4

respectively, and the corresponding values of u by

$$u_0, \qquad u_1, \qquad u_2, \qquad u_3, \qquad u_4;$$

for example, $u_3 = u(x_0 - h, y_0)$. From (3.8.3) we have

$$h^2 \frac{\partial^2 u_0}{\partial x^2} = \delta_x^2 u_0 - \tfrac{1}{12}\delta_x^4 u_0 + \ldots ,$$

where the x-suffix denotes differencing with respect to x keeping y constant. As a first approximation we take

$$h^2 \frac{\partial^2 u_0}{\partial x^2} = \delta_x^2 u_0 = u_1 - 2u_0 + u_3,$$

similarly

$$h^2 \frac{\partial^2 u_0}{\partial y^2} = \delta_y^2 u_0 = u_2 - 2u_0 + u_4,$$

hence

$$h^2 \left(\frac{\partial^2 u_0}{\partial x^2} + \frac{\partial^2 u_0}{\partial y^2}\right) = u_1 + u_2 + u_3 + u_4 - 4u_0$$

and (7.3.1) multiplied by h^2 becomes

$$u_1 + u_2 + u_3 + u_4 - 4u_0 = h^2 f(x_0, y_0). \qquad (7.3.2)$$

We define the residual r at (x_0, y_0) by the equation

$$r = u_1 + u_2 + u_3 + u_4 - 4u_0 - h^2 f_0. \qquad (7.3.3)$$

We begin by guessing the values of u at the internal nodes either

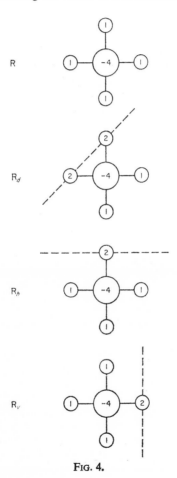

FIG. 4.

arbitrarily or by simple linear interpolation from the boundary values. We apply (7.3.3) to calculate r at each internal point. Next we try to adjust the values of u at the internal points so as to make the residuals as small as possible. For this purpose we consider the effect on r_0, r_1, r_2, r_3, r_4 when we give u_0 an increment of 1. Applying (7.3.3) at the point 0 we see that r_0 decreases by 4 units. Applying (7.3.3) at the point 1 we see that r_1 increases by 1 unit; this is because u_0 relative to the point 0 becomes u_3 relative to the point 1. Similarly we can show that each of r_2, r_3, r_4 increases by 1. This is represented conveniently by the relaxation operator R_0 in Fig. 4. If we apply R_0 at a point next to the boundary, one or more of the end points of R_0 is chopped off because there are no residuals at the boundary. At each step we pick up the node

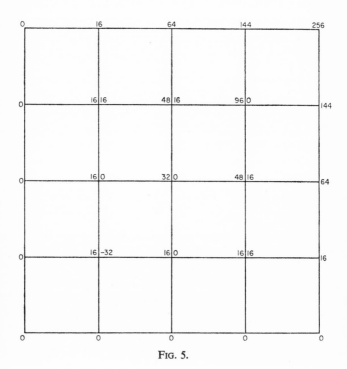

Fig. 5.

with the numerically largest residual and apply to it a suitable multiple of R_0 so as to liquidate the residual as far as possible.

Example. To solve Laplace's equation

$$\frac{\partial^2 u}{\partial x^2} + \frac{\partial^2 u}{\partial y^2} = 0$$

inside the square bounded by the lines $x = 0$, $x = 4$, $y = 0$, $y = 4$ given that $u = x^2 y^2$ on the boundary.

We take $h = 1$ and use the formula $u = x^2 y^2$ to find u on the boundary; we guess u at the internal nodes by simple interpolation and use (7.3.3) to evaluate r at these points; thus we get Fig. 5.

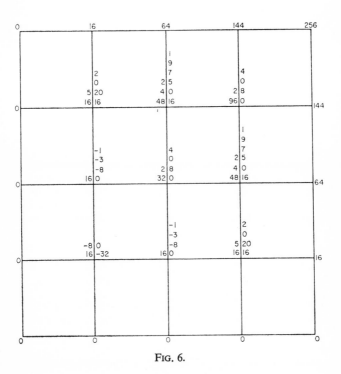

FIG. 6.

200 NUMERICAL ANALYSIS

As in the previous section r is recorded on the right and u on the left.

Applying $-8R_0$ at (1, 1); $4R_0$ at (3, 2) and (2, 3); $5R_0$ at (3, 1) and (1, 3); $2R_0$ at (3, 3); and $2R_0$ at (3, 2) and (2, 3) we get Fig. 6. As in the previous section, there are two stacks of numbers at each node. In the left stack we record the increment to u so that that current value of u is the sum of the left stack. On the right stack we record the modified residual, so that the current value of the residual at each node is the top number of the right stack.

We start afresh in Fig. 7 by recording the final u's and checked r's in Fig. 6. Next we apply $1R_0$ at (2, 2); (3, 3); (3, 2) and (2, 3); (3, 1) and (1, 3); (2, 2); (3, 3); (3, 2) and (2, 3); we obtain Fig. 8. The final solution is in Fig. 9 which is a checked summary

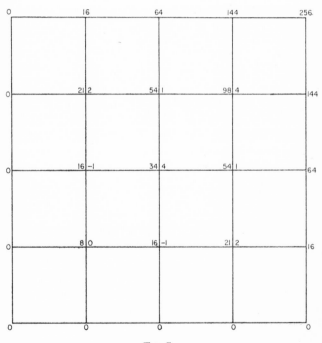

FIG. 7.

of Fig. 8. On a computation sheet only Figs. 6 and 8 need appear, and possibly Fig. 9 for the final answer; the other figures show initial and intermediate stages.

In the following remarks we show how this work can be speeded up considerably and how to solve such problems on digital computers.

Remark 1. Because of the symmetry along the diagonal where $i = j$, we could limit ourselves to the half containing the points (i, j) for which $i \geq j$; this can save a great deal of work. However the relaxation operator R_0 has to be replaced by the operator R_d in Fig. 4 whenever we relax points adjacent to the diagonal, i.e. at the points (2, 1) and (3, 2). The operator R_d gives an

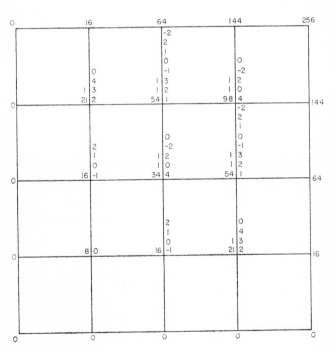

FIG. 8.

increment of 2 units to the residuals at points on the axis of symmetry, 1 unit from each side. Similarly if there is a vertical axis of symmetry in the problem we could save work by computing only the left half, but at points adjacent to the axis of symmetry we must apply the operator R_v shown in Fig. 4; and similarly the operator R_h in Fig. 4 is used for a horizontal axis of symmetry. If there are two or three axes of symmetry the area can be reduced to a quarter or an eighth of the original size.

As in the previous section we introduce two ideas to speed up convergence. We can use block relaxation where the solution is simultaneously displaced by equal amounts at several adjacent points, for example at the 4 corners of a square mesh. Convergence is also speeded up if we make the residuals alternate in

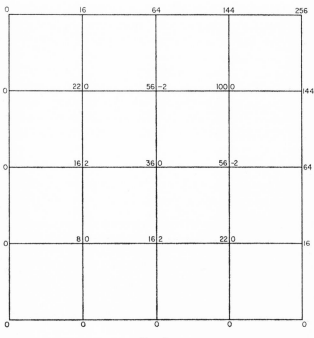

FIG. 9.

sign along consecutive nodes in both directions of the two-dimensional mesh; as in the previous section this method is called over-relaxation.

To improve accuracy we can, as in the previous section, introduce more figures and more nodes, and take correction terms into account. To get reasonable accuracy in actual problems, we frequently need a number of points which is too large to be practical or possible by hand computations. Further, the problem is often more complicated because the boundary is not rectangular and because the partial differential equation is more complicated than Poisson's. We therefore have to consider methods suitable for digital computers.

Remark 2. We can introduce matrix methods as in Remark 1 of the previous section. If, say, we divide each side of the square into $n + 1$ equal intervals we get n^2 internal points. Taking $n = 20$, which is the sort of size one needs in order to get reasonable accuracy, this gives 400 unknown u's and a matrix equation $\mathbf{Ax} = \mathbf{b}$ where $\mathbf{A}$ is (400 × 400). The matrix $\mathbf{A}$ has a band structure similar to but rather more complicated than that in Remark 1 of the previous section. Thus we see that matrices arising from partial differential equations quickly get out of hand and special iterative methods have to be used to solve such matrices. We explain two such methods suitable for digital computers.

The first method is equivalent to the relaxation method of this section except for two differences. On digital computers we need no longer use simple multiples and integers; we can liquidate the residual completely by taking

$$u_0 = \tfrac{1}{4}(u_1 + u_2 + u_3 + u_4 - h^2 f_0).$$

As it is relatively time-consuming to scan a two-dimensional array to pick up the largest residual, we liquidate the residuals in turn in a fixed order; thus we need not form or store the residuals. If this method is expressed in matrix notation, it can be seen that it is equivalent to the Gauss–Seidel method explained in §7.1.

Example. We apply this method to the problem solved above.

Point	1st iteration	5th iteration	9th iteration
(1, 1)	8·000	7·976	8·266
(2, 1)	14·000	16·260	16·552
(3, 1)	19·500	21·844	21·990
(1, 2)	14·000	16·260	16·552
(2, 2)	31·000	35·688	35·980
(3, 2)	52·625	55·272	55·419
(1, 3)	19·500	21·844	16·552
(2, 3)	52·625	55·272	55·419
(3, 3)	98·313	99·636	99·709
Largest residual	8·000	0·312	0·019

This procedure can be speeded up considerably by the method of "over-relaxation". This method has the same effect as the over-relaxation technique explained earlier in this section and the previous section. The Gauss–Seidel method explained above can be put in the form

$$u_0' = u_0 + r,$$

where

u_0' = the new, improved, solution,

u_0 = the previous solution

r = residual = $\frac{1}{4}(u_1 + u_2 + u_3 + u_4 - 4u_0 - h^2 f_0)$.

The method of over-relaxation can be expressed in the form

$$u_0' = u_0 + wr,$$

where u_0', u_0, r are as above. The number w is called the "over-relaxation parameter". If we take $w = 1$ the method becomes the Gauss–Seidel method. For certain other values of w convergence can be much faster. Usually w lies in the range $(1 < w < 2)$. The best value depends upon the problem; when the number of internal nodes is very large then the best w is nearly 2; the theoretical formula for the best w is beyond the scope of this book. We apply this method to the problem solved above for comparison; we take $w = 1·2$.

Point	1st iteration	5th iteration	9th iteration
(1, 1)	6·400	8·282	8·286
(2, 1)	13·120	16·569	16·571
(3, 1)	19·936	22·002	22·000
(1, 2)	13·120	16·569	16·571
(2, 2)	30·272	36·002	36·000
(3, 2)	53·462	55·431	55·429
(1, 3)	19·936	22·002	22·000
(2, 3)	53·462	55·431	55·429
(3, 3)	99·277	99·715	99·714
Largest residual	8·000	0·088	0·0002

EXERCISES

1. Solve the set of equations

$$25x + 2y + z = 70$$
$$2x + 10y + z = 60$$
$$x + y + 4z = 40$$

to 4 decimal places using the Gauss–Seidel iterative process with appropriate checks.   (U.L.)

2. Use relaxation methods to find y when $x = 1, 2, 3, 4, 5, 6, 7$ given that $y = 0$ when $x = 0$ and $x = 8$ and given the following table for y''

x	1	2	3	4	5	6	7
y''	10	4	−2	−8	−14	−20	−26.

(U.L.)

3. A square is bounded by the lines $x = 0$, $y = 0$, $x = 4$, $y = 4$. The temperature V along the boundary is given by

$$V = 128 \text{ along } x = 0$$
$$V = 128 + x^3 \text{ along } y = 0$$
$$V = 192 - 12y^2 \text{ along } x = 4$$
$$V = 128 - 48x + x^3 \text{ along } y = 4,$$

and throughout the region

$$\frac{\partial^2 V}{\partial x^2} + \frac{\partial^2 V}{\partial y^2} = 0$$

Use the method of relaxation to determine the temperature at the 9 internal points $x = m$, $y = n$ for $m, n = 1, 2, 3$.   (B.C.)

   [*Ans.* 126, 130, 146; 117, 112, 119; 102, 82, 74.]

4. Figure 10 represents in scale proportion a square metal plate in which a square hole is punched centrally. The inner boundary is maintained at 100°C and the outer boundary at 0°C, and throughout the plate the temperature satisfies

$$\frac{\partial^2 V}{\partial x^2} + \frac{\partial^2 V}{\partial y^2} = 0.$$

Find to the accuracy of the finite difference approximation the temperature at the upper right-hand corner of each of the squares marked a, b, c, d in the figure. (B.C.)

[*Ans.* 21, 42, 48, 49.]

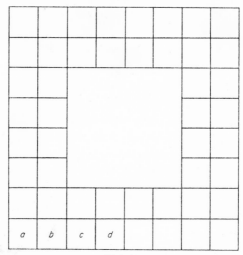

FIG. 10.

5. Solve the differential equation

$$2xy'' + y' + 2xy = 0 \qquad \begin{aligned} y &= 0 \text{ when } x = 0 \\ y &= 2 \text{ when } x = 1 \end{aligned}$$

in the range (0, 1) taking $h = 0 \cdot 1$. (U.L.)

6. Calculate the smallest eigenvalue and the corresponding eigenfunction of the differential system

$$y'' + \lambda x^2 y = 0, \qquad y(0) = y(1) = 0 \qquad \text{(U.L.)}$$

CHAPTER 8

Numerical Methods for Unequal Intervals

8.1 Interpolation

Given the values

$$y_0, \ y_1, \ y_2, \ \ldots \ , \ y_n$$

of a function $f(x)$ at the $n + 1$ distinct points

$$x = x_0, \ x_1, \ x_2, \ \ldots \ , \ x_n$$

we wish to estimate y at some other value x. To do this we approximate $f(x)$ by a polynomial and, in particular, choose one which passes through the points (x_0, y_0), (x_1, y_1), $\ldots$, (x_n, y_n). There is a unique polynomial of degree n through these $n + 1$ points. For if there were two such polynomials, $h(x)$ and $g(x)$ (say) then $h(x) - g(x)$ would be a polynomial of degree n at most with $n + 1$ zeros at $x = x_0, x_1, \ldots , x_n$. But this is impossible, so $h(x) - g(x)$ must be identically zero hence $h(x)$ is unique.

Lagrange's interpolation polynomial is derived as follows. First we define $p(x)$, a polynomial of degree $n + 1$, from the equation

$$p(x) = (x - x_0)(x - x_1)(x - x_2) \ \ldots \ (x - x_n)$$

Next we define $n + 1$ polynomials

$$p_0(x), \ p_1(x), \ p_2(x), \ \ldots \ , \ p_n(x)$$

each of degree n, from the equation

$$p_r(x) = \frac{p(x)}{x - x_r}$$
$$= (x - x_0)(x - x_1) \ \ldots \ (x - x_{r-1})(x - x_{r+1}) \ldots$$
$$\ldots \ (x - x_n).$$

We note that

$$p_r(x_s) = \begin{cases} 0 & \text{when } s \neq r \\ \\ p_r & \text{when } s = r \end{cases}$$

where $\qquad\qquad p_r = p_r(x_r) \neq 0.$

Hence the polynomial

$$(y_r/p_r)\, p_r(x)$$

is a polynomial of degree n such that

$$(y_r/p_r)\, p_r(x) = \begin{cases} 0 & \text{when } x = x_s \neq x_r \\ \\ y_r & \text{when } x = x_r \end{cases}$$

Hence the required interpolation polynomial is the sum of the $n + 1$ polynomials of the above form:

$$L(x) = (y_0/p_0)\, p_0(x) + (y_1/p_1)\, p_1(x) + \\ + \ldots + (y_n/p_n)\, p_n(x) \qquad (8.1.1)$$

Clearly $L(x)$ is the required approximation to $f(x)$; it is a polynomial of degree n at most and is such that

$$L(x_r) = y_r = f(x_r)$$

and further it is the only such polynomial.

Let

$$f(x) = L(x) + R$$

where R is the remainder term which determines how good is the approximation $L(x)$. We will show below that

$$R = p(x) f^{(n+1)}(u)/(n + 1)! \qquad (8.1.2)$$

where $f^{(n+1)}(u)$ is the $(n + 1)$th derivative of $f(x)$ at some point u within an interval containing

$$(x, x_0, x_1, \ldots, x_n).$$

Generally $p(x)$, and so R, is an oscillatory function of x in this range, the amplitude of the oscillation increasing towards the two ends of the range; outside the range, however, R tends to $\pm \infty$ (see Fig. 11). Thus the approximation $L(x)$ gives the best results about the middle of the range and is useless towards the ends of the range or outside the range.

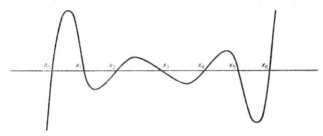

FIG. 11.

In general we have a large number of pairs (x_i, y_i) in the form of a table of, say, 100 entries. To use formula (8.1.1) one takes n about 4 or 5, and selects

$$(x_0, y_0), (x_1, y_1), \ldots , (x_4, y_4)$$

in the neighbourhood of x for which y is required. When f varies slowly n can be less than 5; when f changes rapidly n has to be larger than 5. Formula (8.1.1) has several disadvantages. For a given n, all of the $n + 1$ terms in (8.1.1) have to be calculated as they are of the same order of magnitude; if n has to be made larger then all the calculations have to be repeated; there is no easy estimate of the accuracy of the formula. The next method to be explained, *Newton's interpolation formula with divided differences*, does not have these disadvantages.

First we define and study the properties of divided differences. There are divided differences of order 1, 2, ..., and generally divided differences of order r are defined recursively in terms of divided differences of lower order. We define a first divided difference $f(x_0, x_1)$ thus

$$f(x_0, x_1) = \frac{f(x_1) - f(x_0)}{x_1 - x_0}$$

and similarly we define the first divided difference $f(x_1, x_2)$ thus

$$f(x_1, x_2) = \frac{f(x_2) - f(x_1)}{x_2 - x_1}.$$

We observe that the order of the x's does not matter

$$f(x_0, x_1) = f(x_1, x_0)$$

and that, by the mean value theorem,

$$f(x_0, x_1) = f'(u),$$

where u is some point which lies in the range (x_0, x_1).

Next we define the second divided difference $f(x_0, x_1, x_2)$ thus

$$f(x_0, x_1, x_2) = \frac{f(x_1, x_2) - f(x_0, x_1)}{x_2 - x_0}.$$

And similarly we define $f(x_1, x_2, x_3)$ and observe that the order of the x's is immaterial, and that, as will be shown later,

$$f(x_0, x_1, x_2) = f''(u)/2!,$$

where u lies in an interval containing (x_0, x_1, x_2).

And generally we define the rth divided difference by

$$f(x_0, x_1, \ldots, x_r) = \frac{f(x_1, x_2, \ldots, x_r) - f(x_0, x_1, \ldots, x_{r-1})}{x_r - x_0},$$

and show later that

$$f(x_0, x_1, \ldots, x_r) = f^{(r)}(u)/r!,$$

where u lies in an interval containing $(x_0, x_1, \ldots, x_r)$.

Next we introduce divided differences involving x and $f(x)$; it is immaterial for the definitions that we may not yet know $f(x)$. Thus

$$f(x, x_0) = \frac{f(x) - f(x_0)}{x - x_0},$$

$$f(x, x_0, x_1) = \frac{f(x, x_0) - f(x_0, x_1)}{x - x_1},$$

. . .

$$f(x, x_0, x_1, \ldots, x_r) = \frac{f(x, x_0, \ldots, x_{r-1}) - f(x_0, x_1, \ldots, x_r)}{x - x_r}.$$

Multiplying the above equations by their respective denominators and rearranging we get

$$f(x) = f(x_0) + (x - x_0)f(x, x_0),$$
$$f(x, x_0) = f(x_0, x_1) + (x - x_1)f(x, x_0, x_1),$$
$$\cdots \cdots \cdots \cdots \cdots \cdots \cdots$$
$$f(x, x_0, \ldots, x_{n-1}) = f(x_0, x_1, \ldots, x_n) +$$
$$+ (x - x_n)f(x, x_0, \ldots, x_n).$$

Starting with the last of these $n + 1$ equations, we substitute the left-hand side of each equation in the right-hand side of the previous one. The first equation becomes

$$f(x) = f(x_0) + (x - x_0)f(x_0, x_1) +$$
$$+ (x - x_0)(x - x_1)f(x_0, x_1, x_2) +$$
$$+ \ldots + (x - x_0)(x - x_1)$$
$$\ldots (x - x_{n-1})f(x_0, x_1, \ldots x_n) + R \quad (8.1.3)$$

where

$$R = R(x)$$
$$= (x - x_0)(x - x_1) \ldots (x - x_n)f(x, x_0, x_1, \ldots, x_n).$$

Thus we can put (8.1.3) in the form

$$f(x) = P(x) + R(x)$$

where $P(x)$ is a polynomial of degree n, and since $R(x_i) = 0$ for $i = 0, 1, \ldots, n$, it follows that

$$f(x_i) = P(x_i), \qquad i = 0, 1, \ldots, n.$$

Hence $P(x)$ is the required polynomial approximation to $f(x)$; it follows that $P(x)$ is the same as Lagrange's polynomial $L(x)$ since the interpolation polynomial of order n which agrees with $f(x)$ at $x = x_0, x_1, \ldots, x_n$, is uniquely determined.

We consider now the remainder term R. Since $R(x)$ is zero at $x = x_0, x_1, \ldots, x_n$, it follows, by repeated use of the mean value theorem, that $R^{(n)}(u) = 0$ for some u in the interval containing $x_0, x_1, \ldots, x_n$. Hence

$$f^{(n)}(u) = P^{(n)}(u).$$

$P(x)$ is the sum of $n + 1$ terms which are polynomials of degree $0, 1, 2, \ldots, n$, respectively; hence if we differentiate $P(x)$ n times the first n terms become zero and the derivative of the last term is $n!f(x_0, x_1, \ldots, x_n)$; hence

$$f(x_0, x_1, \ldots, x_n) = f^{(n)}(u)/n!$$

This gives the above-mentioned expressions for divided differences in terms of higher derivatives; these expressions are valid only if $f(x)$ is continuous and its higher derivatives exist up to the required order. If we replace n by $n + 1$ we get

$$f(x, x_0, x_1, \ldots, x_n) = f^{(n+1)}(u)/(n + 1)!,$$

this time u is in the interval containing $(x, x_0, x_1, \ldots, x_n)$. Substituting this value in the expression for $R(x)$ we get (8.1.2).

The formula for the polynomial $P(x)$ in (8.1.3) is called Newton's interpolation formula with divided differences. Most of the interpolation formulae in Chapter 3 are special cases of this polynomial where the points $x_0, x_1, x_2, \ldots$, are taken in a certain order and spacing; hence we can use the formula for the remainder term in (8.1.2). To get the best accuracy when using $P(x)$ we should arrange $x_0, x_1, x_2, \ldots$, in order of increasing distance from x; for example, if y is given at $x = 0\cdot1, 0\cdot3, 0\cdot4, 0\cdot7$, and $0\cdot9$ and we require y at $x = 0\cdot52$, we take

$$x_0 = 0\cdot4, \quad x_1 = 0\cdot7, \quad x_2 = 0\cdot3, \quad x_3 = 0\cdot9, \quad x_4 = 0\cdot1.$$

Normally, and with this arrangement of the x's, the terms in $P(x)$ get smaller and smaller and thus we need only calculate the terms which are significant to a certain specified accuracy.

We now explain *Aitken's method of linear cross means* or successive linear interpolations; this method is equivalent to Newton's formula with divided differences, but the arithmetic is organized in a simple manner so that each step is a simple linear interpolation.

As before, the problem is to find an estimate of $y(x)$ given x and the points

$$(x_0, y_0), (x_1, y_1), (x_2, y_2), \ldots$$

We define a sequence of interpolation polynomials as follows:

$$y_{01}(x) = \frac{1}{x_1 - x_0} \begin{vmatrix} y_0 & x_0 - x \\ y_1 & x_1 - x \end{vmatrix}.$$

We can verify that $y_{01}(x)$ is a polynomial of degree 1 derived by linear interpolation between y_0 and y_1, and that

$$y_{01}(x_0) = y_0, \qquad y_{01}(x_1) = y_1.$$

Similarly we define

$$y_{02}(x) = \frac{1}{x_2 - x_0} \begin{vmatrix} y_0 & x_0 - x \\ y_2 & x_2 - x \end{vmatrix}.$$

Next we define

$$y_{012}(x) = \frac{1}{x_2 - x_1} \begin{vmatrix} y_{01}(x) & x_1 - x \\ y_{02}(x) & x_2 - x \end{vmatrix}.$$

We can verify that $y_{012}(x)$ is a polynomial of degree 2 derived by linear interpolation between two polynomials of degree 1, and that

$$y_{012}(x_i) = y_i \text{ for } i = 0, 1, 2.$$

And generally we define

$$y_{012\ldots n}(x) = \frac{1}{x_n - x_{n-1}} \begin{vmatrix} y_{012\ldots n-1}(x) & x_{n-1} - x \\ y_{01\ldots n-2\,n}(x) & x_n - x \end{vmatrix}$$

and verify that $y_{012\ldots n}(x)$ is a polynomial of degree n derived

8

by linear interpolation between two polynomials of degree $n - 1$, and that

$$y_{012\ldots n}(x_i) = y_i \text{ for } i = 0, 1, 2, \ldots, n.$$

As the interpolation polynomial of order n, which coincides with the given function at $x = x_0, x_1, \ldots, x_n$, is unique, it follows that $y_{012\ldots n}(x)$ is the same as the polynomial of degree n derived by Newton's formula with divided differences.

We show the method of computation in the table below,

where $y_{012\ldots r}(x)$ is denoted by $y_{012\ldots r}$

and $x_r - x$ is denoted by v_r.

x_0	y_0					v_0
x_1	y_1	y_{01}				v_1
x_2	y_2	y_{02}	y_{012}			v_2
x_3	y_3	y_{03}	y_{013}	y_{0123}		v_3
x_4	y_4	y_{04}	y_{014}	y_{0124}	y_{01234}	v_4

All the entries in a line are computed before proceeding to the next line. As an illustration we show how to obtain the line through (x_4, y_4).

$$y_{04} = (y_0 v_4 - y_4 v_0)/(v_4 - v_0)$$

$$y_{014} = (y_{01} v_4 - y_{04} v_1)/(v_4 - v_1)$$

$$y_{0124} = (y_{012} v_4 - y_{014} v_2)/(v_4 - v_2)$$

$$y_{01234} = (y_{0123} v_4 - y_{0124} v_3)/(v_4 - v_3)$$

Here we are making use of the identity

$$x_i - x_j = v_i - v_j$$

to make the computation slightly more convenient.

To get the best accuracy it is important that

$$x_0, x_1, x_2, \ldots$$

be numbered in order of their distance from x. With this ordering of the x_i, the diagonal elements

$$y_0, y_{01}, y_{012}, y_{0123}, y_{01234}, \ldots$$

in the above table are the best estimates of y for their order. We denote these diagonal elements by

$$u_0, u_1, u_2, u_3, u_4, \ldots$$

respectively. This computation can be organized very efficiently on a digital computer so that when we are about to compute the line through x_n, the only numbers which have to be kept from the previous lines are

$$v_0 \; v_1 \; v_2 \; \ldots \; v_{n-1},$$
$$u_0 \; u_1 \; u_2 \; \ldots \; u_{n-1}.$$

As an example we show how to compute the line through x_n on a digital computer:

Set $v_n = x_n - x$ and $u_n = y_n$.

For $j = 0, 1, 2, \ldots, n-1$, replace u_n by $(u_j v_n - u_n v_j)/(v_n - v_j)$ so that u_n takes the values $y_{0n}, y_{01n}, y_{012n}, \ldots, y_{0123 \ldots n}$, successively, ending up with the best polynomial approximation so far in u_n.

Test to see if u_n, u_{n-1} agree to the desired accuracy and accordingly either terminate the computation or proceed to $n + 1$.

Aitken's method, as described above, is very efficient and suitable for digital computers as it requires very little storage space. It has been extended to calculate the derivatives. We will illustrate it by applying it to the problem of inverse interpolation.

We recall the warning given in §3.4 concerning inverse interpolation. If y is a function of x:

$$y = f(x),$$

this may or may not define x as a function of y:

$$x = g(y).$$

Even if x can be expressed as a function of y, this function may not be suitable for interpolation because it changes rapidly in the investigated range; in such a case it is better to carry out inverse interpolation by the iterative method explained in §3.4

and towards the end of §3.7. If, however, x is a well-behaved function of y then inverse interpolation can be carried out easily with the use of interpolation formulae for unequal intervals by simply interchanging x and y.

Example. Given

$$\sin 22° = 0·374604$$
$$\sin 23° = 0·390731$$
$$\sin 24° = 0·406737$$
$$\sin 25° = 0·422618$$
$$\sin 26° = 0·438371$$

find $\sin^{-1}(0·4)$.

We take

$$x_0 = 0·406737 \qquad y_0 = 24$$
$$x_1 = 0·390731 \qquad y_1 = 23$$
$$x_2 = 0·422618 \qquad y_2 = 25$$
$$x_3 = 0·374607 \qquad y_3 = 22$$
$$x_4 = 0·438371 \qquad y_4 = 26$$

and the problem is to find y when $x = 0·4$.

i	x_i	y_i	y_{0i}	y_{01i}	y_{012i}	y_{0123i}	v_i
0	0·406737	24					0·006737
1	0·390731	23	23·5791				−0·009269
2	0·422618	25	23·5758	23·5781			0·022618
3	0·374607	22	23·5806	23·5782	23·5782		−0·025393
4	0·438371	26	23·5741	23·5781	23·5782	23·5782	0·038371

Thus $y = 23·5782$ to 6 significant figures; this is all the accuracy we can get in this case because

$$\delta y = y'(x)\,\delta x \sim \frac{1}{0·016000}\,0·0000005 = 0·00003,$$

which is the error in y due to rounding errors in x.

On a desk machine the computation can be simplified very

much as we can drop any common leading digits. For example,

$$(23 \cdot 5791v_4 - 23 \cdot 5741v_1)/(v_4 - v_1)$$
$$= 23 \cdot 57 + (91v_4 - 41v_1)/10000(v_4 - v_1).$$

8.2 Curve Fitting and the Method of Least Squares

We first make some general remarks concerning the various ways in which a function may be specified and approximated.

A function may be specified in the form of a table:

$$(x_1, y_1), (x_2, y_2), \ldots, (x_n, y_n) \qquad (8.2.1)$$

where n is usually large, say of the order of 100. This is called the *discrete case*.

A function may be specified by a numerical procedure which may be very lengthy but which will determine y, more or less accurately, corresponding to any specified x in the range (a, b):

$$y = f(x) \qquad a \leqq x \leqq b. \qquad (8.2.2)$$

This is called the *continuous case*.

In either case there is a need for some simple and quick procedure which will approximate y for any given x. We distinguish two ways of approximating y.

The first way is typified by the various interpolation formulae of Chapter 3 and §8.1. We have a formula which gives a *local* approximation to the function desired, the actual formula varying as we move from one point to another in the table (8.2.1) or in the range (8.2.2). The resulting values are usually in exact agreement with the given values at the base points; but the accuracy usually deteriorates as the point x moves away from some central position for the particular formula.

The second type of approximation represents the given function *globally*, i.e. over the whole range of x, by a single, fairly simple formula. The overall agreement may not be exceptionally good, but may nevertheless be adequate if we need a single formula to suffice everywhere in the range or if the accuracy of the specified data or function does not warrant high accuracy. In the rest of this chapter we study this type of approximation which is called *curve-fitting*. It is often applied to experimental data when statistical methods of fitting, such as the method of *least squares*, are appropriate.

Let $f^*(x)$ be the required approximation to $f(x)$. Whether the approximation is local or global we usually express $f^*(x)$ in the form

$$f^*(x) = u_0 p_0(x) + u_1 p_1(x) + \ldots + u_m p_m(x). \qquad (8.2.3)$$

Usually m is small, say of the order 10, and the $p_r(x)$ are some simple functions of x. We give some familiar illustrations.

(a) Taylor's series may be considered a form of local approximation where

$$p_r(x) = x^r.$$

(b) Interpolation formulae are local approximations where, for example in the Gregory–Newton formula, we have

$$p_r(x) = \binom{u}{r}, \qquad u = (x - x_0)/h.$$

(c) Fourier series expansions are global approximations where

$$p_{2r}(x) = \cos rx, \; p_{2r+1}(x) = \sin rx, \qquad -\pi \leqslant x \leqslant \pi.$$

In order that the functions $p_r(x)$ should be some simple polynomials, we often associate with the range (a, b) a positive *weight function* $w(x)$. If it helps, the beginner can ignore $w(x)$ and take it to be 1. In the discrete case we associate with the table (8.2.1) positive weight numbers

$$w_1, w_2, \ldots, w_n. \qquad (8.2.4)$$

Again, we can frequently ignore the weight numbers because they are all equal to unity.

All the formulae below are expressed in a form suitable for the continuous case, although in fact they are frequently used for the discrete case. To do this we replace integration by summation.

For example, the expression

$$\int_a^b w(x)\,p(x)\,f(x)\,\mathrm{d}x$$

becomes

$$w_1 y_1 p(x_1) + w_2 y_2 p(x_2) + \ldots + w_n y_n p(x_n)$$

and the expression

$$\int_a^b w(x)\,f^2(x)\,\mathrm{d}x \qquad (8.2.5)$$

becomes $\qquad w_1 y_1^2 + w_2 y_2^2 + \ldots + w_n y_n^2.$

Expressions of the form (8.2.5) occur frequently in the rest of this chapter. We denote the expression (8.2.5) by $N(f)$; it is always positive and its square root is called the *norm* of $f(x)$:

$$\| f(x) \| = [N(f)]^{\frac{1}{2}} = \left[\int_a^b w(x)\,f^2(x)\,\mathrm{d}x \right]^{\frac{1}{2}}. \qquad (8.2.6)$$

It measures, in a sense, the order of magnitude of $f(x)$.

The problem of curve fitting consists in:

(a) The choice of the functions $w(x)$ and $p_r(x)$; we will first assume that they are given, and later we show how to determine them.

(b) The determination of the coefficients

$$(u_0, u_1, \ldots, u_m)$$

in the approximation (8.2.3). These are derived by the method of least squares. The derivation can be simplified if we use matrix notation, and first we quote two matrix results.

(i) If $\mathbf{x}$ is any column vector then the matrix product

$$\mathbf{x}^\mathrm{T}\mathbf{x} = \mathbf{x} \cdot \mathbf{x}$$

is a scalar quantity, the scalar product of the vector $\mathbf{x}$ by itself.

(ii) Given two matrices $\mathbf{A}, \mathbf{B}$ for which the matrix product $\mathbf{AB}$

is defined, then
$$(\mathbf{AB})^{\mathrm{T}} = \mathbf{B}^{\mathrm{T}}\mathbf{A}^{\mathrm{T}}.$$

We return to the problem of curve fitting. Let
$$r(x) = f^*(x) - f(x)$$
be the residual function which measures the accuracy of the approximation $f^*(x)$. Let
$$N(r) = ||\, r(x)\, ||^2 = \int_a^b w(x)\, r^2(x)\, \mathrm{d}x. \qquad (8.2.7)$$

In the discrete case $N(r)$ is the sum of the squares of the residuals for the approximation $f^*(x)$ at the n tabular points, multiplied by the corresponding weights.

If we expand the right-hand side of (8.2.7) using (8.2.3) we can express $N(r)$ as a quadratic in $(u_0, u_1, \ldots, u_m)$. We choose the coefficients $(u_0, u_1, \ldots, u_m)$ so that the quadratic $N(r)$ is minimum; this method is called the principle of least squares. The stationary value of $N(r)$ is obtained by differentiating it with respect to $(u_0, u_1, \ldots, u_m)$ and equating the derivatives to zero; we get a system of $m + 1$ linear equations in the $m + 1$ unknowns $(u_0, u_1, \ldots, u_m)$, called the *system of normal equations*. Since $N(r)$ is positive for any choice of $(u_0, u_1, \ldots, u_m)$, it follows that the stationary value of $N(r)$ corresponding to the solution of the normal system is a minimum. The normal system can be derived using matrix notation as follows. Let
$$\mathbf{A} = (a_{ij})$$
be the $(m + 1 \times m + 1)$ matrix where
$$a_{ij} = \int_b^a w(x)\, p_i(x)\, p_j(x)\, \mathrm{d}x. \qquad (8.2.8)$$

Let
$$\mathbf{u} = (u_0, u_1, \ldots, u_m)^{\mathrm{T}},$$
$$\mathbf{b} = (b_0, b_1, \ldots, b_m)^{\mathrm{T}}$$
where
$$b_i = \int_b^a w(x)\, p_i(x)\, f(x)\, \mathrm{d}x. \qquad (8.2.9)$$

If we use these matrix and vector notations, then expand the right-hand side of (8.2.7) by means of (8.2.3) we get:

$$N(r) = \mathbf{u}^\mathrm{T}\mathbf{A}\mathbf{u} - 2\mathbf{u}^\mathrm{T}\mathbf{b} + N(f). \qquad (8.2.10)$$

This is a simple matrix expression for what would otherwise be a complicated quadratic expression in the u's. Differentiating (8.2.10) with respect to the u's and equating to zero, the normal system of linear equations becomes

$$\mathbf{A}\mathbf{u} = \mathbf{b}. \qquad (8.2.11)$$

If we substitute this value of b in (8.2.10) we get the following expression for the minimum value of $N(r)$:

$$\min N(r) = N(f) - \mathbf{u}^\mathrm{T}\mathbf{A}\mathbf{u}. \qquad (8.2.12)$$

In general the coefficients of $\mathbf{A}$ and $\mathbf{b}$ are difficult to evaluate, and even when they are evaluated the solution of (8.2.11) is often very difficult because the matrix $\mathbf{A}$ is ill-conditioned. The matrix $\mathbf{A}$ is called the *normal matrix*; we observe that from (8.2.8) we have

$$a_{ij} = a_{ji} \text{ hence } \mathbf{A}^\mathrm{T} = \mathbf{A},$$

so that $\mathbf{A}$ is symmetric.

The solution of (8.2.11) is, however, easy if $\mathbf{A}$ is a diagonal matrix so that

$$a_{ij} = a_{ji} = 0 \text{ when } i \neq j,$$

i.e.
$$\int_a^b w(x)\, p_i(x)\, p_j(x)\, \mathrm{d}x = 0 \text{ when } i \neq j. \qquad (8.2.13)$$

The system of functions $p_r(x)$ used in (8.2.3) to approximate $f(x)$, is said to be an *orthogonal system* if it satisfies (8.2.13). For such a system (8.2.11) can be solved immediately:

$$u_i = b_i/N_i,$$

where b_i is defined by (8.2.9) and $N_i = a_{ii}$ is defined by (8.2.8):

$$N_i = \int_a^b w(x)\, p_i^2(x)\, \mathrm{d}x, \qquad (8.2.14)$$

and (8.2.12) becomes:

$$R = \min N(r)$$
$$= N(f) - N_0 u_0^2 - N_1 u_1^2 - \ldots - N_m u_m^2. \qquad (8.2.15)$$

This is a simple but important expression which shows that the minimum residual R decreases when m is increased. R measures the accuracy of the "best fit"; best, that is, in the sense of least squares. The system of functions $p_r(x)$ is said to be *complete* if R tends to zero when m increases. For example, the system of functions

$$p_r(x) = \cos rx$$

is incomplete for the range $(-\pi, \pi)$ since we cannot approximate functions which are not even in terms of even functions only.

The same method and notation can be used to solve a different problem; the solution of an over-determined system of equations:

$$\mathbf{Px} = \mathbf{p},$$

representing a linear system of m equations in n unknowns where m is bigger, usually much bigger, than n. Usually there is no solution which satisfies the given system. But we seek a solution which minimizes, in the sense of least squares, the residual vector

$$\mathbf{r} = \mathbf{Px} - \mathbf{p};$$

then

$$R = \mathbf{r} \cdot \mathbf{r} = \mathbf{r}^T \mathbf{r} = \mathbf{x}^T \mathbf{A} \mathbf{x} - 2\mathbf{x}^T \mathbf{b} + \mathbf{p}^T \mathbf{p}$$

where

$\mathbf{A} = \mathbf{P}^T \mathbf{P}$ is the normal symmetric $(n \times n)$ matrix,

$\mathbf{b} = \mathbf{P}^T \mathbf{p}$.

The quantity R is a positive quadratic in the components of $\mathbf{x}$; by differentiating it with respect to these components and equating the derivatives to zero, we get the normal system of equations

$$\mathbf{Ax} = \mathbf{b}.$$

The solution of this system is the required best solution of the given over-determined system. If we substitute this value of $\mathbf{x}$ in the expression for R we get

$$\min R = \mathbf{p}^T\mathbf{p} - \mathbf{x}^T\mathbf{A}\mathbf{x},$$

much as in (8.2.12).

8.3 Orthogonal Polynomials

Two functions $f(x)$, $g(x)$ are said to be *orthogonal* in the range (a, b) with respect to the weight function $w(x)$ if

$$\int_a^b w(x)f(x)g(x)\,\mathrm{d}x = 0$$

It was shown in the previous section that orthogonal functions are very useful in curve fitting. In this section we consider the case when these functions are polynomials

$$p_0(x),\ p_1(x),\ p_2(x),\ \ldots, \tag{8.3.1}$$

are polynomials of degree 0, 1, 2, ..., respectively. For simplicity we make the leading coefficients (i.e. the coefficients of the highest power) of the polynomials $p_i(x)$ unity.

In the first part of this section we discuss properties which hold in general. Although the formulae are expressed in a form suitable for the continuous case (8.2.2) we emphasize here again that the results apply also to the discrete case (8.2.1), then the corresponding formulae are derived by replacing integration by summation as in (8.2.5). In fact some of the techniques explained in the first part of this section are more suitable and more frequently used for the discrete case when we are required to fit curves to experimental data. In the second part of this section we give some classical orthogonal polynomials frequently used in applications for approximations in the continuous case.

Given any polynomial $P(x)$ of order n, we can express it, by successive polynomial divisions, in the form

$$P(x) = u_np_n(x) + u_{n-1}p_{n-1}(x) + \ldots + u_0p_0(x) \tag{8.3.2}$$

where $u_n, u_{n-1}, \ldots, u_0$ are constants and $u_n \neq 0$.

Hence

$$\int_a^b w(x)\, p_i(x)\, P(x)\, \mathrm{d}x = 0 \ \text{if}\ i > n \qquad (8.3.3)$$

i.e. $p_i(x)$ is orthogonal to all polynomials of degree lower than i.

We use L_i, N_i to denote the following expressions which occur frequently in this section:

$$N_i = N[p_i(x)] = \int_a^b w(x)\, p_i^2(x)\, \mathrm{d}x, \qquad (8.3.4)$$

$$L_i = \int_a^b w(x)\, x p_i^2(x)\, \mathrm{d}x. \qquad (8.3.5)$$

We observe, as in (8.2.5), that

$$N_i = ||\, p_i(x)\, ||^2 > 0.$$

We seek a recurrence relation to express $p_{i+1}(x)$ in terms of

$$p_i(x),\ p_{i-1}(x),\ \ldots,\ p_0(x).$$

Let

$$xp_i(x) = A_0 p_{i+1}(x) + A_1 p_i(x) + \\ + A_2 p_{i-1}(x) + \ldots + A_i p_0(x). \quad (8.3.6)$$

By equating the coefficients of x^{i+1} we get $A_0 = 1$. If we multiply the two sides of (8.3.6) by $w(x)\, p_r(x)$ and integrate over the range (a, b) we get the following results for various values of r. For $r = i + 1$ we get

$$\int_a^b w(x)\, xp_i(x)\, p_{i+1}(x)\, \mathrm{d}x = N_{i+1};$$

hence, replacing i by $i - 1$, we get

$$\int_a^b w(x)\, xp_{i-1}(x)\, p_i(x)\, \mathrm{d}x = N_i. \qquad (8.3.7)$$

For $r = i$ we get

$$L_i = A_1 N_i.$$

For $r = i - 1$ we get, using (8.3.7),

$$N_i = A_2 N_{i-1}.$$

For $r < i - 1$ we get

$$A_3 = A_4 = \ldots = A_n = 0.$$

Hence (8.3.6) can be rearranged in the form of a recurrence relation:

$$p_{i+1}(x) = (x - a_i) p_i(x) - b_i p_{i-1}(x), \qquad (8.3.8)$$

where

$$a_i = L_i/N_i, \qquad b_i = N_i/N_{i-1}. \qquad (8.3.9)$$

The recurrence relation (8.3.8) is frequently used to evaluate the polynomials $p_r(x)$, provided we know the coefficients a_i, b_i, and provided we start with

$$r = 0, \qquad p_0(x) = 1, \qquad b_0 = 0,$$

so that $p_{-1}(x)$ need not be defined.

Orthogonal polynomials are used mainly in order to compute unknown or complicated functions

$$f(x) = u_0 p_0(x) + u_1 p_1(x) + u_2 p_2(x) + \ldots . \qquad (8.3.10)$$

from their approximation

$$f^*(x) = u_0 p_0(x) + u_1 p_1(x) + \ldots + u_m p_m(x). \qquad (8.3.11)$$

Usually m is small, of the order 10. The approximation (8.3.11) would be useless if it were difficult to evaluate the coefficients u_i or if it were difficult to evaluate f^* for any specified x. We now show how these two problems can be solved by numerical procedures which are particularly suitable for fitting curves to experimental data using digital computers.

Given a large set of data

$$(x_1, y_1), (x_2, y_2), \ldots , (x_n, y_n), \qquad (8.3.12)$$

and corresponding weights

$$w_1, w_2, \ldots , w_n$$

where n is large, say of the order 100, we proceed as follows:

(a) Use (8.3.4) and (8.3.5) to calculate L_i, N_i, and (8.3.9) to calculate a_i, b_i.

(b) Calculate u_i from the formula

$$u_i = \int\limits_a^b w(x)\, p_i(x)\, f(x)\, \mathrm{d}x / N_i \qquad (8.3.13)$$

derived by multiplying the two sides of (8.3.10) by $w(x)\, p_i(x)$ and integrating (i.e. summing over the tabular points).

From a_i, b_i we can compute $p_{i+1}(x)$ using the recurrence relation (8.3.8). So that now we can repeat (a) and (b) with i replaced by $i + 1$. Although n is large, usually m is small and (a), (b) are repeated $m + 1$ times for $i = 0, 1, \ldots , m$, ending up with the following $3m$ quantities

$$\left.\begin{array}{c} a_0, a_1, \ldots , a_{m-1}, \\[4pt] b_1, \ldots , b_{m-1}, \\[4pt] u_0, u_1, \ldots , u_{m-1}, u_m. \end{array}\right\} \qquad (8.3.14)$$

The computation of these $3m$ quantities may require a lot of arithmetic, but once they are computed and preserved we can forget about the large amount of data in (8.3.12) and use (8.3.11) to compute the y corresponding to any x, by the following algorithm:

We set

$$S_{m+1} = S_{m+2} = 0$$

and calculate

$$S_m, S_{m-1}, \ldots , S_0$$

by the backward recurrence relation

$$S_r = (x - a_r)\, S_{r+1} - b_{r+1} S_{r+2} + u_r \qquad (8.3.15)$$

for $\qquad r = m, m - 1, \ldots\ 2, 1, 0.$

Then

$$S_0 = y = f^*(x)$$
$$= u_0 p_0(x) + u_1 p_1(x) + \ldots + u_m p_m(x). \qquad (8.3.16)$$

This important result is derived as follows. From (8.3.15) we have

$$u_r = S_r - (x - a_r) S_{r+1} - b_{r+2} S_{r+2}.$$

On substituting this expression for $r = 0, 1, 2, \ldots, m$ in the left-hand side of (8.3.16) we get a long expression; using the recurrence relation (8.3.8) for $r = 0, 1, \ldots, m - 1$, all the terms in this long expression cancel except S_0.

We consider now some useful properties of the zeros of orthogonal polynomials. Let $p_n(x)$ be the $(n + 1)$th polynomial in the orthogonal set (8.3.1); as it is of degree n it has n zeros; we will show that these zeros are real, distinct, and lie inside the range (a, b); i.e. $p_n(x)$ changes sign n times in the range (a, b). For suppose it changes sign r times only at

$$x_1, x_2, \ldots, x_r, \text{ where } r < n$$

and let

$$P(x) = (x - x_1)(x - x_2) \ldots (x - x_r),$$

so that $P(x)$ is a polynomial of degree r which is lower than the degree of $p_n(x)$; it follows from (8.3.3) that

$$\int_a^b w(x)\, p_n(x)\, P(x)\, \mathrm{d}x = 0.$$

Since $w(x)$ is positive, this is only possible if $p_n(x)\, P(x)$ is positive in some parts of the range and negative in other parts of the range. On the other hand the polynomial $p_n(x)\, P(x)$ cannot change sign in the range because all its real roots, including $x_1, x_2, \ldots, x_r$, are of even multiplicity; this leads to a contradiction.

This and many other properties of the roots of the orthogonal set of polynomials (8.3.1) can be derived from the recurrence relation (8.3.8) and the fact that

$$b_i = N_i/N_{i-1} > 0.$$

For example, it follows immediately from (8.3.8) that $p_{i+1}(x)$ and $p_{i-1}(x)$ have opposite signs at the zeros of $p_i(x)$, hence the zeros of $p_i(x)$ and $p_{i+1}(x)$ "interlace", i.e. any two consecutive roots of $p_{i+1}(x)$ are separated by prcisely one root of $p_i(x)$. Consider the signs of the sequense of $n + 1$ numbers

$$p_0(t), p_1(t), \ldots, p_n(t),$$

and let $v(t)$ denote the number of changes of sign in this sequence, then it follows from the property just established that $v(t)$ is equal to the number of roots of $p_n(x)$ lying to the right of t. These properties can be used to compute the roots of $p_n(x)$.

As n becomes large $p_n(x)$ oscillates rapidly in the range of integration (or summation for the discrete case); it follows that

$$u_i = \int_a^b w(x)\, p_i(x) f(x)\, \mathrm{d}x / N_i$$

frequently becomes very small for large i as a lot of cancellation normally takes place in the summation, provided $f(x)$ is a function which does not itself oscillate rapidly. Thus in the approximation

$$f^*(x) = u_0 p_0(x) + u_1 p_1(x) + \ldots + u_m p_m(x)$$

we expect the coefficients u_i to become numerically very small as i gets larger and the first neglected coefficient, u_{m+1}, is usually a reasonable measure of the accuracy of the approximation.

We now list some of the more commonly used orthogonal polynomials in the continuous case. We emphasize that all the properties explained so far apply to each of the sets of special polynomials listed below with a few modifications to allow for the fact that leading coefficients are not unity. In each case we give explicit formulae for the terms a, b, $w(x)$ which occur in the defining relation

$$\int_a^b w(x)\, p_i(x) p_j(x)\, \mathrm{d}x = 0 \text{ when } i \neq j.$$

We give the appropriate recurrence relation (8.3.8) and a formula for

$$N_n = \int_a^b w(x)\, p_n^2(x)\, \mathrm{d}x$$

Finally we list in each case the first six polynomials. For further details consult the references mentioned in the suggestions for further reading.

Legendre polynomials $P_n(x)$

$$a = -1 \qquad b = 1 \qquad w(x) = 1$$

$$N_n = \frac{2}{2n + 1}$$

leading coefficient $= \dfrac{(2n)!}{2^n(n!)^2}$

$$P_{n+1}(x) = \frac{2n + 1}{n + 1} xP_n(x) - \frac{n}{n + 1} P_{n-1}(x)$$

$$P_0(x) = 1 \qquad P_1(x) = x$$
$$P_2(x) = \tfrac{1}{2}(3x^2 - 1)$$
$$P_3(x) = \tfrac{1}{2}(5x^3 - 3x)$$
$$P_4(x) = \tfrac{1}{8}(35x^4 - 30x^2 + 3)$$
$$P_5(x) = \tfrac{1}{8}(63x^5 - 70x^3 + 15x)$$

Chebyshev polynomials $T_n(x)$

$$a = -1 \qquad b = 1 \qquad w(x) = (1 - x^2)^{-\frac{1}{2}}$$
$N_n = \pi$ when $n = 0$, and $\tfrac{1}{2}\pi$ otherwise.
leading coefficient $= 2^{n-1}$ for $n > 0$
$$T_{n+1}(x) = 2xT_n(x) - T_{n-1}(x)$$
$$T_0(x) = 1 \qquad T_1(x) = x$$
$$T_2(x) = 2x^2 - 1$$
$$T_3(x) = 4x^3 - 3x$$
$$T_4(x) = 8x^4 - 8x^2 + 1$$
$$T_5(x) = 16x^5 - 20x^3 + 5x$$

See the second remark below for the applications of these polynomials.

Hermite polynomials $H_n(x)$

$$a = -\infty \qquad b = +\infty \qquad w(x) = e^{-x^2}$$

$$N_n = 2^n n! \, \pi^{\frac{1}{2}}$$

leading coefficient $= 2^n$

$$H_{n+1}(x) = 2xH_n(x) - 2nH_{n-1}(x)$$

$$H_0(x) = 1 \qquad H_1(x) = 2x$$

$$H_2(x) = 4x^2 - 2$$

$$H_3(x) = 8x^3 - 12x$$

$$H_4(x) = 16x^4 - 48x^2 + 12$$

$$H_5(x) = 32x^5 - 160x^3 + 120x$$

Laguerre polynomials $L_n(x)$

$$a = 0 \qquad b = \infty \qquad w(x) = e^{-x}$$

$$N_n = (n!)^2$$

leading coefficient $= (-1)^n$

$$L_{n+1}(x) = (1 + 2n - x) L_n(x) - n^2 L_{n-1}(x)$$

$$L_0(x) = 1 \qquad L_1(x) = 1 - x$$

$$L_2(x) = 2 - 4x + x^2$$

$$L_3(x) = 6 - 18x + 9x^2 - x^3$$

$$L_4(x) = 24 - 96x + 72x^2 - 16x^3 + x^4$$

$$L_5(x) = 120 - 600x + 600x^2 - 200x^3 + 25x^4 - x^5$$

Remark 1. Consider, for example, the problem of approximating
$$F(t) \text{ in the range } a \leqq t \leqq b$$

by Legendre polynomials. The range for Legendre polynomials
is $(-1, 1)$ and not (a, b). This is allowed for by a simple change

of the independent variable:

$$t = \tfrac{1}{2}(a + b) - \tfrac{1}{2}(b - a)\, x$$

so that as t varies from a to b, x varies from -1 to 1. Let $f(x)$ be the function obtained by substituting this expression for t in $F(t)$. Now we can find an approximation

$$f^*(x) = u_0 P_0(x) + u_1 P_1(x) + \ldots + u_m P_m(x)$$

for $f(x)$ which can be evaluated for any t by taking

$$x = \frac{t - \tfrac{1}{2}(a + b)}{\tfrac{1}{2}(b - a)}\,.$$

Remark 2. Chebyshev polynomials have important properties which make them particularly suitable for approximating functions on digital computers. Functions such as trigonometric, exponential, or Bessel functions are frequently used; the method of evaluating them on a computer usually amounts to evaluating a polynomial obtained by truncating a series. It is therefore important that this series should be rapidly convergent so that we need as few terms as possible. For example, to evaluate e^x in the range $(-1, 1)$ from the series

$$e^x = 1 + x + x^2/2! + x^3/3! + \ldots$$

would require ten terms in order to achieve accuracy to 5 decimal places, whereas we need 8 terms to achieve the same accuracy if we expand in terms of Chebyshev polynomials. This is generally true for all the elementary and indeed most functions: Chebyshev series converge faster than other forms of approximations. We will also show that evaluating a Chebyshev series is very little longer than evaluating a polynomial; it can be shown also that the numerical procedure for evaluating a Chebyshev series is usually more accurate, i.e. less affected by error build-up, than evaluating the corresponding polynomials in powers of x.

Consider the approximation

$$f^*(x) = u_0 T_0(x) + u_1 T_1(x) + \ldots + u_n T_n(x). \quad -1 \leqq x \leqq 1$$

From (8.3.13) we get

$$u_r = \frac{2}{\pi} \int_{-1}^{1} (1 - x^2)^{-\frac{1}{2}} f(x) T_r(x) \, dx, \text{ for } r = 1, 2, \ldots, \quad (8.3.17)$$

$$u_0 = \frac{1}{\pi} \int_{-1}^{1} (1 - x^2)^{-\frac{1}{2}} f(x) T_0(x) \, dx.$$

To avoid this exception for $r = 0$ we put the expansion in the form

$$f^*(x) = \tfrac{1}{2} u_0 T_0(x) + u_1 T_1(x) + \ldots + u_n T_n(x) \quad (8.3.18)$$

then (8.3.17) applies to $r = 0$ as well.

Consider the transformation

$$x = \cos \theta$$

then
$$w(x) = 1/\sin \theta.$$

The recurrence relation for Chebyshev polynomials gives

$$T_0(x) = 1$$
$$T_1(x) = \cos \theta$$

and generally

$$T_r(x) = \cos r\theta$$
$$T_{r+1}(x) = 2 \cos \theta T_r(x) - T_{r-1}(x)$$
$$= \cos (r + 1) \, \theta.$$

This shows that $T_n(x)$ does not exceed unity numerically in the range $(-1, 1)$. In terms of θ, the equation (8.3.17) becomes

$$u_r = \frac{2}{\pi} \int_{0}^{\pi} f(\cos \theta) \cos r\theta \, d\theta \text{ for } r = 0, 1, 2, \ldots$$

The series (8.3.18) can be evaluated for any x in much the same way as in (8.3.15) for the general orthogonal expansion, except that in this case we have

$$a_r = 0 \qquad b_r = 1$$

and x has to be replaced by $2x$ for $r > 0$, and, finally, u_0 has to be replaced by $\frac{1}{2}u_0$. Instead, we can derive the following algorithm for evaluating (8.3.18) in much the same way as (8.3.15). Set

$$S_{n+1} = S_{n+2} = 0,$$

then calculate

$$S_n, S_{n-1}, \ldots, S_2, S_1, S_0$$

by the following backward recurrence relation:

$$S_r = 2xS_{r+1} - S_{r+2} + u_r \tag{8.3.19}$$

for $r = n, n-1, \ldots, 2, 1, 0$. Then

$$f^*(x) = \tfrac{1}{2}(S_0 - S_2).$$

Example. To evaluate e^x when $x = 0 \cdot 9$, given the coefficients u_r in the second column below for the expansion of e^x in the range $(-1, 1)$ in the form (8.3.18).

We apply the recurrence relation (8.3.19) with $x = 0 \cdot 9$, and so get the third column below:

r	u_r	S_r
6	0·00004	0·00004
5	0·00054	0·00061
4	0·00547	0·00653
3	0·04434	0·05548
2	0·27150	0·36483
1	1·13032	1·73153
0	2·53213	5·28405

$$\tfrac{1}{2}(S_0 - S_2) = 2 \cdot 45961.$$

The correct value to 6 decimal places is $2 \cdot 459603$.

Remark 3. We see from the trigonometric expressions for $T_r(x)$ and u_r that the Chebyshev series for $f(x)$ is equivalent to the Fourier cosine series expansion of the function $f(\cos \theta)$ which is an even function of θ.

8.4 Integration

We describe in this section integration methods which make use of the properties of orthogonal polynomials; the subject is called *Gaussian quadrature*.

The integration formulae of §3.9 approximates the integral

$$\int_a^b F(x)\, dx \qquad (8.4.1)$$

by an expression of the form

$$A_1 F(x_1) + A_2 F(x_2) + \ldots + A_n F(x_n), \qquad (8.4.2)$$

where the points $(x_1, x_2, \ldots, x_n)$ are equidistant and the coefficients $(A_1, A_2, \ldots, A_n)$ are chosen so as to make the approximation error as small as possible. If we expand the remainder R in powers of h, the fixed interval, we can determine the coefficients A_i so that the remainder does not contain terms in

$$1, h, h^2, \ldots, h^{n-1}.$$

R can then be expressed in the form

$$R = K h^n F^{(n)}(u), \qquad a \leqq u \leqq b \qquad (8.4.3)$$

so that $(8.4.2)$ is exact if $F(x)$ is a polynomial of order not exceeding $n-1$ since the nth derivative $F^{(n)}(u)$ is zero.

If now we relax the condition that the points $(x_1, x_2, \ldots, x_n)$ be equidistant we can expect to raise the order of accuracy by n so that

$$R = k F^{(2n)}(u)$$

and $(8.4.2)$ becomes exact for polynomials of order not exceeding $2n-1$.

The problem will be stated in a slightly more general form. Given a positive weight function $w(x)$ we can define a function $f(x)$ so that

$$F(x) = w(x) f(x).$$

We seek coefficients
$$A_1, A_2, \ldots, A_n \tag{8.4.4}$$

and points
$$x_1, x_2, \ldots, x_n \tag{8.4.5}$$

so that the approximation

$$\int_a^b w(x)\ f(x)\ \mathrm{d}x \sim A_1 w(x_1) f(x_1) + A_2 w(x_2) f(x_2) + \\ + \ldots + A_n w(x_n) f(x_n) \tag{8.4.6}$$

is exact when $f(x)$ is any polynomial of degree not exceeding $2n - 1$.

Let
$$p_0(x), p_1(x), p_2(x), \ldots, p_n(x), \ldots \tag{8.4.7}$$

be the system of orthogonal polynomials defined by the weight function $w(x)$ and the range (a, b) as in the previous section. If we take
$$f(x) = P(x)\, p_n(x)$$

where $P(x)$ is any polynomial of degree $n - 1$, it follows by (8.3.3) that the right-hand side of (8.4.6) is zero independently of the choice of $P(x)$, hence

$$p_n(x_1) = p_n(x_2) = \ldots = p_n(x_n) = 0.$$

Thus the required points $(x_1, x_2, \ldots, x_n)$ are the zeros of $p_n(x)$; the n zeros are real and distinct as shown in the previous section.

To determine the coefficients A_i we take

$$f(x) = q^{(i)}(x) = p_n(x)/(x - x_i). \tag{8.4.8}$$

The polynomial $q^{(i)}(x)$ is of degree $n - 1$ such that

$$q^{(i)}(x_j) = 0 \text{ when } j \neq i,$$

$$q^{(i)}(x_i) \neq 0.$$

Substituting this form of $f(x)$ in (8.4.6) we get

$$A_i = \int_a^b w(x)\, q^{(i)}(x)\ \mathrm{d}x / [w(x_i)\, q^{(i)}(x_i)]. \tag{8.4.9}$$

By making use of the recurrence relations (8.3.8) it can be shown, though not easily, that (8.4.9) is equivalent to the following formula which is more suitable for evaluating A_i:

$$A_i = N_{n-1}/[q^{(i)}(x_i) \, p_{n-1}(x_i)]$$

where N_{n-1} is defined by (8.3.4).

Example. To evaluate the integral

$$I = \int\limits_{2}^{8} \frac{1}{t} \, dt$$

we use the Legendre polynomials defined in the previous section for which $w(x) = 1$. The transformation $t = 5 + 3x$ gives:

$$I = 3 \int\limits_{-1}^{1} \frac{1}{5 + 3x} \, dx.$$

Taking $n = 5$, (8.4.6) becomes

$$I \sim 3(A_1 y_1 + A_2 y_2 + A_3 y_3 + A_4 y_4 + A_5 y_5) \text{ where } y_i = \frac{1}{5 + 3x_i}.$$

The points x_i and the coefficients A_i have been extensively tabulated for various values of n; from tables we have

i	1	2	3	4	5
x_i	−0·906180	−0·538469	0	0·538469	0·906180
A_i	0·236927	0·478629	0·568889	0·478629	0·236927
y_i	0·438316	0·295451	0·2	0·151162	0·129558

This gives $I \sim 1 \cdot 386260$; the exact integral is ln $4 = 1 \cdot 386294$. To get a similar accuracy by Simpson's rule we need about 11 evaluations of y.

We observe that the coefficients A_i in this example are positive; it was pointed out at the end of §3.9 that this property is important because it improves the numerical accuracy of integration formulae. We now show that this property is true in Gaussian quadrature. We choose $f(x)$ to be a polynomial of degree $2n - 2$ of the form

$$f(x) = [q^{(i)}(x)]^2$$

so that (8.4.6) is exact, and, since $f(x_j) = 0$ when $j \neq i$, (8.4.6) gives

$$A_i = \int_a^b w(x)[q^{(i)}(x)]^2 \, \mathrm{d}x / \{[q^{(i)}(x_i)]^2 \, w(x_i)\}.$$

The integral and the denominator are both positive, hence A_i is also positive.

The points x_i and the coefficients A_i have been tabulated for the more useful classical orthogonal polynomials. (See Kopal, *Numerical Analysis*, Chapman & Hall, 1961, 562–581 for a useful selection of these coefficients. See the references in the reading list for more detailed tables.) When the limit of integration is finite at both ends we use Legendre polynomials; we may apply (8.4.6) to the whole range or divide the range into smaller parts and apply (8.4.6) to each part of the range. If one or both limits are infinite we can use Legendre polynomials for the middle part of the range and Laguerre polynomials for the infinite parts of the range. If the integrand has a singularity in the range then we have to use some special methods for integrating over a small neighbourhood of the singularity; the methods of this section and of §3.9 do not apply if there is a singularity in the range.

EXERCISES

1. If
$$f(x) = a_0 + a_1x + a_2x^2 + a_3x^3 + a_4x^4 + a_5x^5$$
find A_1, A_2, A_3 and x_1, x_2, x_3 so that
$$\frac{1}{2} \int_{-1}^{1} f(x) \, \mathrm{d}x = A_1f(x_1) + A_2f(x_2) + A_3f(x_3).$$

Hence evaluate approximately

$$\int_0^{0\cdot5} \frac{1}{1 + x^4}\, dx.$$

Check the accuracy of the result by evaluating the integral from a power series expansion.   (B.C.)

2. Prove the recurrence relation for Chebyshev polynomials,

$$T_{n+1}(x) - 2xT_n(x) + T_{n-1}(x) = 0,$$

for $n \geq 1$, where

$$T_n(x) = \cos\,(n\,\cos^{-1} x).$$

The function $f(x)$ is approximated by the Chebyshev expansion

$$f(x) = c_0T_0 + c_1T_1(x) + \ldots + c_nT_n(x).$$

If the constants

$$d_0, d_1, d_2, \ldots, d_{n+2}$$

satisfy the recurrence relations

$$d_k - 2xd_{k+1} + d_{k+2} = c_k, \qquad d_{n+1} = d_{n+2} = 0,$$

show that $$f(x) = d_0 - xd_1.$$

Discuss the usefulness of this result in computing $f(x)$.   (U.L.)

3. Let

$$w(x) = [x\,(1 - x)]^{\frac{1}{2}}, \qquad a = 0, \qquad b = 1.$$

Construct the first four functions

$$p_0(x) \qquad p_1(x) \qquad p_2(x) \qquad p_3(x)$$

of the orthogonal set with respect to these limits and weight functions. Use these functions to approximate $\sin \pi x$ in this range, determining the coefficients and the value of the integral of the weighted squared error over this range.   (U.L.)

4. Show that the usual Chebyshev polynomials $T_n(x)$ in the range $(-1, 1)$ satisfy the relations

$$T_{n+1}(x) - 2xT_n(x) + T_{n-1}(x) = 0$$

$$T_n(x)\, T_m(x) = \tfrac{1}{2}[T_{n+m}(x) + T_{n-m}(x)]$$

Assuming that $1/(1 + x^2)$ is approximated by a Chebyshev polynomial up to the term in $T_4(x)$ inclusive, find the numerical coefficients of the polynomial and show that the remainder term is

$$- T_6(x)/[99\,(1 + x^2)]$$

Hence estimate the value of the integral

$$\int_{-1}^{1} \frac{1}{1 + x^2} \, dx$$

giving upper and lower bounds. (U.L.)

[*Hint.* If

$$1/(1 + x^2) = u_0 T_0 + u_1 T_1(x) + \ldots + u_4 T_4(x) + R$$

multiply through by $1 + x^2$, use the relation $1 + x^2 = \frac{1}{2}[3T_0 + T_2(x)]$ and the above identity.]

Index